Steve Elsworth lives in London. His previous books
have been for teaching English to foreigners. He is a
journalist and is currently working on environmental and
ecological issues.

TO ̄E
DISPOSED
BY
AUTHORITY

To Elaine Walker

Steve Elsworth

Acid rain

Pluto Press

London and Sydney

First published in 1984 by Pluto Press Limited,
The Works, 105a Torriano Avenue, London NW5 2RX
and Pluto Press Australia Limited, PO Box 199, Leichhardt,
New South Wales 2040, Australia

Cover designed by Clive Challis A.Gr.R.
Diagrams and illustrations by John Chesterman

Set by Grassroots, Kilburn, London NW6
Printed in Great Britain by Photobooks (Bristol) Limited
Bound by W.H. Ware & Sons Limited
Tweed Road, Clevedon, Avon

British Library Cataloguing in Publication Data
Elsworth, Steve
    Acid rain
    1. Acid precipitation (Meteorology)—
    Environmental aspects—Europe
    I. Title
    363.7'394   TD196.A25

ISBN 0 86104 791 5

Contents

Introduction

Just over thirty years ago, London experienced its worst peacetime disaster of the twentieth century. Four thousand people died, two thousand were hospitalised, and an uncounted number were distressingly ill. Animals choked to death, thousands of citizens were confined to their homes for a period of four days, and transport in the capital came to an almost complete standstill.

The reason for this was a familiar and traditional one – smog. December 1952 was a cold month, and London's schools, hospitals, factories and homes were burning vast quantities of coal and oil. On the morning of 5 December, there was an atmospheric inversion – a layer of cold air squatted over the city, trapping the rising smoke, which concentrated in ever greater amounts at ground and rooftop level.

The result was a massive accumulation in the air of black dust and sulphur dioxide – a corrosive gas released when coal or oil is burnt – which blotted out the sun, made it impossible for people to see more than a yard in front of their faces, and destroyed the normal functionings of a capital city.

'The fog was so thick that you couldn't park your car,' said one observer. 'Your passenger had to get out and stand at the kerb with a torch before you could even think about leaving the white line in the middle of the road.' People were injuring themselves so frequently by walking into the lamp posts in Whitehall that the authorities tied flares to the posts at eye-level. The Automobile Association's breakdown vans found it impossible to locate members who telephoned in for help; there was not a half-mile of road in central London where visibility was more than five yards. The main streets were eventually cleared of all traffic apart from the occasional convoy of buses trailing each other nose-to-tail back to their depots. Eight seamen lost their ship, and asked a Port of London policeman to show them the way back to it; five minutes later, he fell into the Thames, closely followed by two of the sailors. The Port of London police went on duty wearing life-jackets and carrying shepherds' crooks for pulling people out of the water. The Smithfield Show was ruined as the cattle succumbed to the effects of the pollution: 160 had to have veterinary treatment, and 13 died.

In short, it was a catastrophe, though treated at the time as merely a discomfort; and it was only towards the end of the month, when the mortality figures began to be compiled, that the true extent of the disaster was appreciated.

This had been a particularly poisonous smog, containing sulphur fumes of a strength never before recorded in London. The effects on people's health were horrendous, and they died literally in their thousands—from respiratory diseases, disorders of the heart and circulatory system, and gastro-enteritis. The elderly, the very young, and those in poor health were particularly at risk.

What has all this to do with acid rain? There are two connections. Firstly, sulphur dioxide, a major contributor to the smog, is the central constituent of what is now called acid rain, and it is produced in the air by the same processes which caused the London smogs. Secondly, the measures taken to alleviate London's smogs did not eradicate air pollution but merely transformed it. Instead of having short-term local effects, sulphur dioxide poisoning was transferred to a longer time-scale and began to affect a wider area.

In the immediate aftermath of the 1952 smog, the government was faced with pressing calls for action and the result was the Clean Air Act 1956. Large areas of London were declared smokeless zones in which coal was forbidden for domestic use; factories and public utilities were ordered to clean the black dust from their smoke; and those industries which produced vast quantities of fumes—notably power stations—were instructed to build tall chimneys in order to carry exhaust gases away from the immediate vicinity.

The Clean Air Act was regarded as immensely successful, and other countries followed the UK's lead. Tall chimneys became an essential part of several clean air policies, and local smogs — though still occurring in certain cities, notably Berlin and Athens in 1983/4 — no longer had the catastrophic effects that had been experienced in London. There was one problem, however, which at the time went unnoticed but has since been accorded increasing importance: the production of sulphur dioxide continued unabated, piped into the atmosphere by thousands of factories and industries throughout Europe. Millions of tonnes of this powerful pollutant are sent into the air every year. The irony is, if the Clean Air Act had not been implemented and the London smogs had persisted, then public pressure would have forced the government to do what was then, and still is, technologically possible — clean the sulphur dioxide, as well as the dust, out of industrial smoke.

Other European countries might then have joined the UK in restricting sulphur output, instead of piping the pollutant 400 feet into the air in the belief that it thus became harmless. If they had done this — and they did not — we would not now be witnessing the destruction of large parts of Europe's environment by acidic pollution. Acid rain is, in that respect, the legacy of the London smogs.

1. Acid words

Defining acid rain

'Acid rain' is a particularly inefficient way of describing the atmospheric pollution currently threatening Europe and North America. The phrase conjures up thoughts of burning rainfall, smouldering blotches on trees, and shredded clothes on washing lines. The reality is, of course, different. The damage is done, not by a sudden killer storm, but by the gradual wearing down of the environment caused by rain that is more acidic than it should be. The effect is one of destabilisation of natural processes rather than instant destruction.

The use of 'rain' is misleading, as the pollution occurs in snow, hail, fog, gas clouds, mist, and dry dust; and the word 'acid' is not entirely helpful as most rainfall is naturally acid anyway, and has been for hundreds of years. But the phrase is useful as an umbrella term to describe a number of different pollutant processes which all arise from the same source, and it will be used as such in this book. Occasionally these processes are referred to as 'acid deposition', which is a more accurate term.

Most people know how to find out if a liquid is acid or not: dip a piece of litmus paper into it and see if the paper changes colour. Finding out *how* acid it is, though, can be more difficult.

The pH scale (short for Paper-Hydrion) is used for measuring acidity and alkalinity. It ranges from 0 to 14, with 14 being the most alkaline, 7 neutral (e.g. distilled water), and numbers below 7 indicating increasing acidity. The scale is logarithmic, so pH 5 is ten times more acidic than pH 6, pH 4 a hundred times stronger than pH 6, pH 3 a thousand times more so, and so on.

Because of interaction with carbon dioxide in the Earth's atmosphere, natural rainfall is already acidic, and in Europe can occur at varying pH levels. Traditionally, acid rain is defined as that with a pH below 5.6, which is the acidity of pure water in combination with carbon dioxide. But it is now accepted that the lower limit of natural acidity can occasionally drop to pH 5.1 or 5.0.[1]

The most acid rain in the UK was recorded in Pitlochry, Scotland, on 20 April 1974.[2] It had a pH of 2.4 — similar to that of lemon juice, and a thousand times more acidic than 'normal' rainfall at pH 5.6. Britain's rain is not usually as acid as this, but it has become 10 − 70 times more acidic than 'normal' rain (pH 4.1 to 4.7),[3] and this has already had deleterious effects on the environment.

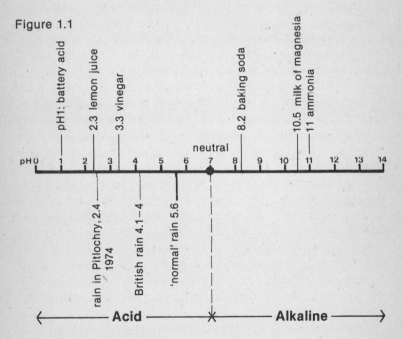

Figure 1.1

The causes of acid rain

Acid rain is a consequence of the combustion of coal and oil on an enormous scale, either for the production of electricity or in certain industrial processes. The by-products from this massive conflagration of fossil fuels are released untreated into the air, and eventually they return to Earth and distort natural cycles.

The two central constituents of the pollution are sulphur dioxide (SO_2) and the oxides of nitrogen (NO_x); SO_2 is generally reckoned to contribute 70 per cent of acid rain, and NO_x 30 per cent.[4]

Sulphur dioxide is produced naturally in considerable quantities — from mud-flats, marshes, swamps, volcanoes, sea spray, and rotting organic material. It forms during the burning of coal and oil because both fuels contain a certain amount of sulphur and this combines with oxygen and rises into the air as sulphur dioxide gas. At the same time, the combustion transforms nitrogen in the fuel and from the surrounding air into nitrogen oxides, also gases.

Before people started producing sulphur dioxide by making fires, it is reasonable to assume that nearly all the SO_2 in the atmosphere was produced from natural processes. Now it is estimated that half the SO_2 is naturally produced, and the other half created by human activity[5] — a significant change in atmospheric loading. Over Europe and eastern North America, this imbalance is even more exaggerated: 90 per cent of the sulphur dioxide in the atmosphere has been put there by us, and only 10 per cent has natural origins.[6]

Although small-scale episodes of pollution have occurred over the centuries, it was the industrial revolution, running as it did on large quantities of coal and later oil, that started this disequilibrium in the air. Initially the pollution was local, centring round heavily-industrialised towns: rain on the outskirts of Manchester and Leeds has contained strong acids for more than a century, and some of the most acid moorland lakes and pools in the UK occur in the southern Pennines, near Huddersfield and Sheffield.[7]

As industrial processes became more widespread, however, the amount of SO_2 entering the atmosphere increased. At the turn of the century, 10 million tonnes (1 tonne (metric) = 0.98 UK ton) of SO_2 per year were being pumped into the air over Western Europe. This increased steadily until the recession period of the

1930s, and was about 12 million tonnes per year by 1950. Over the next twenty years, oil consumption figures leaped, and SO_2 emissions jumped to an annual total of 25 million tonnes by 1970.[8]

Pollution on this scale is almost impossible to imagine. The amount of sulphur in 25 million tonnes of sulphur dioxide is about 12.5 million tonnes, or enough to fill 400,000 thirty-tonne lorries – and all this being dumped onto the lakes and forests of Europe every year. The process has occurred over a period of at least eighty years, with a doubling of annual pollution from 1950 onwards.

Production of nitrogen oxides, though lower in quantity than that of SO_2, has increased by the same proportions. 'NO_x' is short-hand for two gases – nitric oxide (NO) and nitrogen dioxide (NO_2). It is currently thought that Western Europe produces 9 million tonnes of NO_x per year,[9] though this is difficult to estimate. SO_2 calculations are based on consumption of coal and oil with an estimate of 90 per cent of the sulphur content being released as SO_2; it is not possible to do this for NO_x, as some of it is formed from the air during the combustion process. It is generally agreed, however, that creation of artificial NO_x doubled between 1959 and 1973, and is still increasing.[10]

The doubling of SO_2 and NO_x production over the last thirty years has been paralleled by increasing acidity in rainfall. While records of rain acidity prior to the 1960s are not always available, where they can be obtained the results are alarming. A report commissioned by the EEC[11] looked at present rainfall acidity in the UK, the Netherlands, Germany, Sweden, Norway, and Italy, and found that it ranged between pH 3.97 and 4.9, with one reading of 5.5 (in Italy). Such figures as were available for the 1950s showed a range of pH 4.5 to 6, indicating that the rain in the areas examined had become 10 – 80 times more acid in about thirty years.

It was also in the fifties and sixties that increasing acidic effects began to be observed in the lakes in Scandinavia. Appendix A covers the history of observations and research into acidification, but it will suffice to say here that the damage caused to the environment by air pollution made itself slowly apparent up to the 1950s, when it started to become rapidly more visible. It accelerated through the 1960s and 1970s, and now in the mid-eighties has reached crisis proportions, affecting or threatening 2−4 million square miles of Europe and North America.[12] Although the pollution, having industrial causes, is centred in these two areas, it has also been reported in Brazil, South Africa, India, Turkey, Thailand, China−in short, anywhere that is undergoing rapid industrialisation.

How it works

The pollution process can best be understood by thinking of it in two phases, origin and destination. Figure 1.2 shows the general outline of this process. The pollutants are emitted from a large town, with all the industries, electricity needs, and traffic that occur in a normal urban centre. Some of the houses burn coal, others have oil-fired central heating−systems that are also used by local schools and hospitals. There is a coal-burning power station to meet the city's electricity needs situated some distance outside the city limits. It is a winter afternoon, so there is a demand for heat and light: boilers, furnaces, fires, internal combustion engines, all are burning fossil fuels and emitting smoke.

The fumes from all this activity are released into the air above and around the town. The smoke from the cars and houses is released quite close to the ground and does not rise very much. Some of it attaches itself to buildings, drops into gardens, or is caught by the wind and lifted over the suburbs, falling on to nearby fields and woods. A small amount is picked up by eddying wind currents and is mixed with the emissions from the power station and the industrial chimneys. This smoke is about to hitch a long-distance ride: the chimneys are 500 feet tall, and are designed so that the majority of fumes they emit are lifted by currents in the upper air and wafted away from the town.

Sulphur dioxide can stay in the air for up to four days and, if carried by a strong wind, can travel hundreds of miles in that time−sometimes well over a thousand. Depending on the wind currents, for example, pollution emitted over the UK can get to Scandinavia in two or three days. During this journey, the pollutants sometimes combine with water in the clouds to form dilute sulphuric

Figure 1.2. Diagram showing the formation of acid deposition. The pollutants can return to earth as mists, fog, snow, sleet, hail, or as dry particles of sulphates and nitrates.[13]

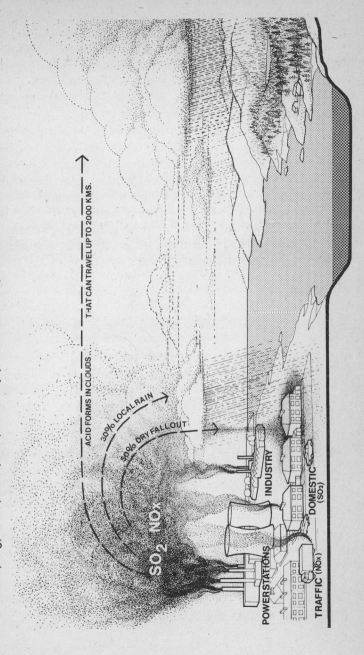

and nitric acids (H_2SO_4 and HNO_3) which then fall to the ground in rain, hail, snow, or sleet. The acids are sometimes neutralised by substances such as ammonia in the air or split into their component parts—in the case of sulphuric acid into hydrogen (H^+) and sulphate (SO_4^{2-}); and nitric acid into hydrogen (H^+) and nitrate (NO_3^-)—which also drop to the ground with environmental consequences similar to those of the falling acids.

There are different stages of pollution. When the smoke is emitted, some of it is transferred to the immediate environment with no interference from water in the atmosphere. This is referred to as 'dry' deposition, which has 'direct' effects. Some pollutants are transported through the air, where they combine with water before returning to earth in a variety of forms. This is 'wet' deposition, and the effects it has on the environment are often said to be 'indirect'. Although 'indirect' effects *sound* less damaging than 'direct', this is not always the case.

Effects

Starting in the town, near the source of pollution:

- Sulphur oxides attack buildings, corroding limestone, sandstone, marble, concrete, and mortar. They also shorten the working life of metals and paints, and have a perishing effect on materials such as leather, textiles, and paper. Historic buildings and works of art are particularly at risk.
- The dry deposition falls on fields outside the town. High concentrations of pollution can create a wilderness in which it it impossible to grow anything except stunted grass. Lower levels reduce the output of a wide range of crops, and damage trees.
- Sulphur oxides in the air mix with fog and dust causing corrosive smogs with lethal results, as in the London 1952 smog. At lesser levels of intensity, this air pollution provokes illness among the elderly and those with respiratory problems. It is also suspected of contributing to lung cancer, and is currently being investigated in

Germany as a possible contributor to Sudden Infant Death Syndrome ('cot deaths').

Significant though the above effects are—according to a 1982 report[14] Europe is losing at least $3,000 million (£2,000 million[15]) a year in corrosion costs alone—they have not been responsible for the present European-wide controversy over acid deposition. What is arousing entire communities and sparking furious diplomatic arguments are the indirect effects of the pollution:

- The sulphur and nitrogen compounds travel some distance through the air, crossing into countries far away from the source of origin before falling to the ground. Over the long period of time in which this has been occurring, there have been certain environmental consequences. Trees in the coniferous forests of Europe initially manage to withstand the poisonous input, but as the acids build up over time, the forests' defence systems collapse, and the trees start to die.
- The pollutants drain into the earth and release poisonous metals such as aluminium, cadmium, and mercury from their compounds in the soil. Nutrients such as calcium or magnesium are depleted by the forests' defence mechanisms against acidity, or are also released to run into the nearest water base. The trees, still fighting against the continual acid input, lose their vitality with the depletion of their stocks of nutrients and are prey to attack from ozone pollution, fungus, insects, disease, or drought.
- The acids and metals move to the nearest lake or stream, which has already been receiving extra acidity direct from rain. Unless it has a way of countering the acidity, this surface water becomes steadily more acidic. The combination of low pH and increasing concentration of metals (notably aluminium) affects most varieties of fish until they eventually die out. Other aquatic species die too, leaving a lake with a very limited variety of life forms that provides too barren an environment for fish to

survive. Such a lake is dead to all intents and purposes, even if it provides a comfortable habitat for certain plants and plankton that actually like acid water. The mobilisation of poisonous metals and the collapse of the lake ecosystem, threaten long-term consequences for species which do not live in the lake but are dependent upon it for survival. One of these consequences is potential danger to human health.

The process is inexorable and progressive, currently affecting thousands of lakes, millions of trees, and billions of fish. Following chapters examine each stage of the pollution in detail, and chapter 4 documents, on a country-by-country basis, the alarming extent of acidic damage. A statement made in June 1983 by the director of the United Nations Environment Programme provides the overall context:[16]

> In Northern Europe, Canada, and the north-eastern United States, the rain is turning rivers, lakes, and ponds acidic, killing fish and decimating other water life. It assaults buildings and water pipes with corrosion that costs millions of dollars every year. It may even threaten human health, mainly by contaminating drinking water. It is a particularly modern, post-industrial form of ruination, and is as widespread and careless of its victims, and of international boundaries, as the wind that disperses it.

There is another aspect to acid rain that needs to be considered in conjunction with its geographical spread, and that is time. The indications are that this type of pollution builds up to a threshold level before triggering environmental destruction, and the increasing reports of damage caused by air pollution seem to show the threshold being triggered more and more rapidly over a widening range of locations. The corrective measures needed to prevent the emission of sulphur and nitrogen oxides will take about ten years to implement — and that will be *after* the countries concerned agree on their implementation, a point which has not yet been reached. Given the accelerating pace of environmental degradation at the moment, a ten-year delay on emission abatement could have extremely wide and severe consequences. The present visible effects of acidification are serious and disturbing, and cost the EEC member countries, according to one report, at least £33,000 million a year in corrosion and environmental destruction.[17] This, however, is only the hors d'oeuvre; the main meal is yet to come.

2. Acid effects

Forests

A forest is not just a collection of trees, passively standing in the sunshine; it is a powerful, living organism, driven by direct and indirect solar energy which feeds ecosystem processes like photosynthesis, nutrient uptake, nitrogen fixation, and growth. The quantities of power used by these ecosystems are enormous: in New England USA, for example, the forests consume 30 times the amount of energy used by human beings living in the area.[1]

These woodlands are breeding places for thousands of species of plants, animals and microbes, which exist alongside hundreds of organic and inorganic compounds all linked together by biological and physical processes. They are not static, but ever-changing as populations wax and wane in response to various disturbances.

There was a time when trees covered a much larger proportion of Europe then they do now. With the growth of farming and the expansion of the population, more land was taken over for arable purposes, and the area of forest land shrank. In many regions, the trees retreated to marginal land – thin soil that was no good for farming purposes, the sides of mountains, and areas where the ground was already quite acidic.

Another change has taken place, too. With the realisation that forestry could be an economic proposition, the species of trees have been changed over the years. As an example, oaks may take 150 years to grow and then yield 4 − 6 cubic metres of wood per hectare (1 hectare = 10,000 square metres or about 2.5 acres). Conifers require only 50 years to mature and can produce 18 cubic metres of wood per hectare. So, in the UK, we plant conifers — which already account for more than 60 per cent of Britain's 2 million hectares of woodland. In fact, the Forestry Commission regards the financial imperative as being so important that in 1976 and 1977 it poisoned 1,000 hectares of young oaks; and according to the Centre of Agricultural Study, 30 conifers are now being planted for every broadleaf (deciduous) tree.[2]

When an acid-bearing cloud arrives over a forest it affects the trees according to their position, their species, and their height above sea level. In hilly country, those trees at a higher altitude will receive more rain than those below — and sometimes a cloud will sit on a hill, enveloping everything in an acidic mist. Similarly, trees exposed to the wind, or taller than their neighbours, will receive a disproportionate dose of pollution. When acidification takes a grip, the vegetation does not all perish simultaneously: some will die, while the rest continues to appear perfectly healthy, on the surface at least.

Various trees are more sensitive than others to air pollution: evergreens are much more susceptible than deciduous trees, probably because the latter get rid of their leaves in the autumn, thus shedding a large part of the acidic input. The former retain their greenery throughout the year and suffer an accumulation of acidity. In fact, current scientific belief is that coniferous trees increase local acidification by attracting pollutants from the air and channelling them into the soil in more acid concentrations.[3] This factor is particularly ironic given the widespread preference for growing coniferous trees on economic grounds. Our choice of tree species has actually helped to make the problem worse.

The 'canopy' is the roof of the forest — those leaves and parts of trees directly exposed to the sky, which shelter everything lower down. Between 20 and 50 per cent of acid precipitation is buffered — or neutralised — by the forest canopy. When the acid falls on to the leaves, the tree summons up alkaline reserves, and a chemical process takes place on the surface of the leaf which counters the effects of the acidity. The tree acquires these reserves by absorbing them from the soil via its roots.[4] If you remove alkalinity from a substance, you make it more acid; so in the initial stages of acidification, the tree protects its upper surfaces at the expense of acidifying the soil.

Some of the pollution falls on the trunk or the branches of the tree in sulphate or nitrate form (or, in cases of local pollution, as SO_2 and NO_x) to be washed into the soil by subsequent rainfall. Other precipitation will fall directly onto the soil, which in time will build up its acidity. The acid water moving through the soil will wash out nutrients, and release poisonous metals — copper, aluminium, cadmium, etc. — from their compounds. Some of these are absorbed by the tree roots.

No single occurrence is responsible for the death of a tree as the process of pollution does not happen overnight. It can take decades, twenty or thirty years — a long struggle for survival against an ever-increasingly acidic diet. Forests are accustomed to dealing with a certain amount of acid input: in the natural order of events, they create acidity in the soil, redressing the balance when the trees eventually die and rot. The forest is a balanced ecosystem, putting back into the environment what it has taken out. The modern form of overloading is, however, often too much for the trees to accommodate. A hot year occurs, finally relieved by rainfall creating an 'acid push' in the ground (making it much more acid for a short period of time) which attacks and kills some of the tree roots. After that, the weakened organism starts to give up the battle, and can be killed by any combination of outside pressures — drought, beetles, fungi, an attack of ozone or other air pollutant, or even frost. Its needles change colour and start to drop, and finally the result is a dead tree.

Or, to be precise, not just one dead tree. If the poisoning of the air persists, other trees will succumb too. In extreme cases, everything in the area will eventually die. At the EEC symposium on acid deposition in Karlsruhe, 1983, Dr Bormann of Yale University presented an analysis of the different stages of forest death, with the consequences incurred by the ecosystem. The following is a shortened version of his findings.[5]

Stage 0
Pollutant levels insignificant. Pristine ecosystem.
Stage 1
Pollutant occurs at generally low levels. Species and ecosystem functions relatively unaffected.

Stage 2a

Level of pollutants hostile to some aspects of the life cycle of sensitive species or individuals. For example, some species may suffer reduced photosynthesis, a change in reproductive capacity, liability to insect or fungus attack, or change in nutrient cycling.

Stage 2b

Increased pollution stress. Populations of sensitive species decline. Ultimately these species may be lost from the system, but a more likely fate is that some individuals will hang on. If a plant is an important member of an ecosystem, its loss can have reverberating effects. For example, the successful growth of seedlings of one species may depend on environmental conditions, like shade, provided by another. If the second species is sensitive to pollution, the first may suffer also. There are also non-plant complications involving other members of the system. Honeybees are particularly sensitive to air pollution, and conditions deleterious to their survival might interfere with the pollination of plants, seed production, and ultimately with the re-establishment of vegetation.

Stage 3a

More pollution stress. The size of plants becomes important to survival: large trees, shrubs, and growths of all species die. The basic structure of the ecosytem is changed. The forest is now reduced to small scattered shrubs and herbs, including weedy species not previously present. The probability of fire increases with the extra amount of dead wood lying around. The capacity to filter pollutants from rainwater is lowered, and more of these substances are transferred directly to streams and rivers.

Stage 3b

Ecosystem collapse. The degeneration of the system has been described as 'peeling off the forest structure'. First the trees perish, followed by tall shurbs, and finally, under the severest conditions, the short shrubs and herbs succumb. This kind of collapse is, of course, not unknown in natural conditions: all ecosystems encounter various kinds of disturbances such as fires, storms, outbreaks of insect pests, or large-scale felling of trees. In the course of several decades, the ecosystem will rebuild itself. However, as Dr Bormann writes:

On the other hand, some disturbances can so damage ecosystems through loss of species, ecosystem structure, nutrients and soil, that the capacity of the ecosystem to repair itself is greatly restricted. Even if the perturbing force is removed, *these damaged ecosystems would take centuries or even millenia to achieve predisturbance levels of productivity, structure, and function.* It is this type of degraded ecosystem that is found under the severest pollution conditions close to point sources. [emphasis added]

Dr Bormann is referring to 'point pollution sources' — that is, heavy local deposition such as that encountered near a large industrial plant pouring out SO_2 and NO_x. Much of the previous part of this section is based on the work of Professor Ulrich, of Göttingen University, Germany, who is not solely concerned with trees near these 'point sources', and who argues that there is a *general* and progressive increase in the number of tree deaths in Germany. Ecosystems have not collapsed as dramatically as described by Dr Bormann, but that is not to say that future collapse is impossible.

Forest damage in the Federal Republic of Germany is bad enough at the moment. According to the Research and Development Section of the EEC Commission,[6] 2 million hectares (7,700 square miles) of coniferous forest show some signs of acidification — that is over half of the coniferous forests of Germany. In addition 560,000 hectares (2,000 square miles) are classified as 'completely devastated areas'. Bavaria has 160,000 hectares (600 square miles) of affected woodland, Baden-Wurttemberg 130,000 hectares (500 square miles), North Rhine-Westphalia over 70,000 hectares (270 square miles). On some estimates the resulting loss of production will wipe out 47,000 jobs in the German forestry and wood-processing industry.

The speed with which air pollution damage is spreading has taken scientific observers by surprise. In 1982, a national survey in West Germany found that 7.7 per cent of the country's trees were affected. As a result of this, a more thorough investigation was launched in 1983, which discovered that the proportion of blighted trees was actually 34 per cent, with one in ten suffering 'medium or severe damage'. Overall, 41 per cent of spruce trees were affected, 75 per cent of firs, 26 per cent of beech trees and 15 per cent of oaks.[7] The devastation of the fir trees means that there is a real risk of whole forests being destroyed. 'If pollution continues at the present rate, most of the fir and spruce trees in the Black Forest will be dead by the 1990's,' says Gerhard Weiser, agriculture and forestry minister for Baden-Wurttemberg.[8]

The EEC is seriously concerned because one-third of the Community's agricultural land is under forest; there is a major threat, not just to environmental protection, but to the economy as a whole. The Community's forest products sector employs 1.4 million people, which puts it in the same league as the motor, chemicals, and textiles industries.[9]

Apart from Germany's 560,000 hectares (2,000 square miles) of completely devastated woodland, 390,000 hectares (1,500 square miles) of forest in Poland have been affected, 490,000 hectares (1,900 square miles) in Yugoslavia, and 100,000 hectares (380 square miles) in Czechoslovakia. Forest areas are also suffering damage in the German Democratic Republic, the Netherlands, Switzerland, France and Austria. Fuller details are given in chapter 4 under the country headings.

It is very difficult to estimate the precise costs of these forest deaths, partly because of the time-scales involved and partly as a result of different interpretations of the size of the diseased areas. According to some estimates, the value of the damage throughout the EEC is £57 million per year and rising.[10] According to others, the costs to German forests alone is £250 million annually.[11] The European parliament points to an estimated German capital loss of 10,000 million DM (£2,500 million) and an annual loss of 400 million DM (£103 million).[12] Until a precise figure is agreed, it will be difficult to use the financial implications of widespread tree deaths as an argument in favour of massive 'desulphurisation' programmes to remove the sulphur from industrial smoke.

'Using the financial implications' is an impressive phrase which has echoed through the scientific and diplomatic discussions on whether to desulphurise or not. This reaction to the problem of the forest deaths — to try to find out how much it will cost — is of course part of the process that started the damage in the first place. Pollutants pour out of power stations, factories, and large public buildings because when they were built it was decided on economic grounds not to install smoke-cleaning machinery. The effect of acidity is exacerbated in the countryside by falling on coniferous trees which have been planted for their financial profitability with little or no thought for the ecological consequences. And now the people who prompted this debacle — the accountants and economists — are sent in to try and sort out the mess.

There is a sort of wooden stupidity about a cost-benefit analysis of a forest death. Environmentalists watch with open mouths as the statisticians go round with their notebooks, counting the numbers of dead trees. They can see the calculations going down on the paper: 'You get £20-worth of wood from this tree, and there are about a million trees in this forest, so this investment is worth, let me see, £20 million.' End of calculation. It is a Kafkaesque process, working under the rationale that a forest is worth the total sale price of its wood: the difference between the Black Forest and a useless, infertile, plain is the number of telegraph poles that each can produce.

This approach begs two questions. First, how do you put a price on the Black Forest? Is, for example, a 2 per cent loss per year merely another entry in the economic ledger, or an irreversible ecological catastrophe which could last for centuries and should be avoided at all costs? How do you cost the effect of a dead forest on the quality of life of the local community?

The second question refers to the long-term possibilities raised by continuous air pollution. Trees and plants survive through photosynthesis — a process by which, using sunlight as their energy source, they convert carbon dioxide (CO_2) and water into sugar and oxygen. We depend on the vegetation of the Earth to replenish our oxygen supplies and forests provide a large part of that replenishment.

Air pollution, according to the framework of ecosystem damage that Bormann has described, reduces photosynthesis and does this at a relatively early stage in the pollution chain. Moreover, according to the United Nations Economic Commission for Europe, there is evidence to suggest that tree growth may be decreased in air which contains an annual average of $25-50$ micrograms of SO_2 per cubic metre (1 microgram is one millionth of a gram). This kind of air is found 'over large parts of Europe'.[13]

So there is a possibility—and no one has been able to test this yet—that trees in large areas of Europe are not growing as much as they should; and also, where significant pollution occurs, the rate of photosynthesis is reduced. Does this mean that the overall rate of photosynthesis in Europe is decreasing? Have the economists calculated the cost-effect if photosynthesis was reduced by, say, 10 per cent because of these two pollution-linked changes? And, importantly, if the fixation of carbon dioxide by plants is decreased by 10 per cent does this mean that the 10 per cent would now be surplus and build up in the atmosphere?

It is an important question because the forecasts of what would happen in the event of a CO_2 overload in the atmosphere are much worse than the situation we are encountering now with SO_2 and NO_x pollution. Some climatologists hypothesise that discharge into the atmosphere of CO_2 and other pollutants is likely to sufficiently change the radiation balance of the Earth within the next century or so, to cause major warming as well as changes in the patterns of climate. There is considerable uncertainty in the predictions but some of the hypothesised consequences are extremely severe. Droughts may be caused, capable of reducing food production in many areas. The changes in climate would also produce changes in many of the Earth's natural ecosystems. The rise in sea level caused by the break-up or melting of the Greenland and west Antarctic ice sheets might necessitate the eventual relocation of major cities.[14]

This is all hypothetical as yet, but it is framed by evidence of a dramatically increasing amount of CO_2 present in the atmosphere. How can these possibilities be linked to a 'cost-benefit' analysis of tree deaths in Bavaria? They cannot—because we do not train our economists to take the long-term ecological consequences into account when they are making their calculations. If they did, then maybe we would not be faced, as we are now, with the widespread destruction of the forests of central Europe.

Lakes and streams

Not all experts are agreed about the air pollution damage to the forests of Europe. There is still a heated scientific discussion as to the extent and exact causes of the phenomenon. This argument is to be expected: the theory of pollution-related tree deaths only became developed in the late 1970s, and the life cycle of most trees is so long that disease symptoms, particularly those occurring over a large area as the result of a slow process of poisoning, take some time to become apparent. After that there has to be a delay while the scientists collect information about what is happening; and following this there is further delay while the exact meanings of the different data are argued over and finally assessed.

The destruction of aquatic life is much more easily quantified, however. New fish generations are produced every year and can be measured annually. When the environment in which they live changes radically, the results very quickly become apparent. If this ecological change occurs over a large area it will not escape notice, firstly through the reports of anglers and professional fishing people, and secondly through the work of scientists and water authority specialists following up the initial observations. It may take several years to find out if a forest is dying as a result of air pollution: the pH of a lake can be tested in an afternoon.

The disappearance of fish from lakes and streams has been recorded in Scandinavia since the beginning of the century, but especially over the last twenty years. Lakes have 'died' in many parts of Sweden, Norway, Canada and the USA, and increasingly acid waters have been reported in Denmark, the Netherlands, West Germany, Switzerland, and the UK. 'Increasingly acid' is a term which needs explaining, and Lake Gårdsjön in Sweden can be used as an example. The sediment of this lake was studied to discover its pH levels from 12,500 BC to the present time. During this period the water acidity remained between pH 6 and pH 7 (i.e. almost neutral) for about 14,000 years, until the early 1960s. Since that time it has fallen from pH 6 to pH 4.5, i.e. it has become 70 times more acid than it had been previously.[15] This is not to say that all acidified watercourses have dropped 1.5 units—but they have experienced uncharacteristic increases in acidification.

In the same way that the pollution of the forests is not a straightforward simple process, the acidification of surface waters is complex and subtle. It is not possible to say that every lake in a certain area is going to become acidified by a given time, because so much depends on the type of soil that surrounds the lake, the local geology, and the size and depth of the water body. What can be said with certainty, however, is that the acidification of lake systems has multiplied over the last twenty years, and shows every sign of continuing to do so. According to a 1981 report from the USA National Research Council: 'At current rates of emission of sulphur and nitrogen oxides, the number of affected lakes can be expected to more than double by 1990, and to include larger and deeper lakes.'[16] If nothing is done to halt this process, we can expect the acidification of our lakes and streams to continue apace.

There are various ways in which the increased acidic input gets into the lake ecosystem. It can fall directly on to the lake surface as wet deposition (rain, snow, etc.), as dry dust, or as gas. Each type of input affects the acidity of the lakewater to some extent. If the lake bottom consists of material containing sandstone or limestone, this acts as a 'buffering' agent and neutralises the incoming acid. The result is that the pH of the water remains the same with no short-term environmental upsets. If, on the other hand, the lake bottom is composed of a hard rock such as granite or gneiss, then this defence system is not available and the water starts to acidify.

The other entry point for acidity into the watercourse is via the catchment area surrounding the lake. Acid rain falls on to trees, crops and soil, and eventually drains into the nearest waterbase, being filtered through the soil on its way there. There are a number of buffering agents in the soil that can blunt the effects of acidity before the water reaches its destination and much depends on these buffering agents. If they do not perform efficiently then the acid rainfall is transferred directly into lakes and streams.

Again, the results of overloading the environment begin to be observed. The rainfall is not overwhelmingly acidic, but there is a lot of it. The buffering capacities of various ecosystems are fairly fixed: some can withstand acid inputs for long periods of time, others for much shorter timespans. In the latter type, the buffering agents in the soil and the lake bottom become progressively weaker under the relentless pollutant input, until eventually they collapse. The pH of the lakewater begins to drop; poisonous metals, either carried into the lake with the acidified rainfall or released from the lake bottom, begin to circulate in the water and the inhabitants of the ecosystem start to die.

The organisms do not all perish at the same time, and individuals can continue to survive after most of their class has disappeared, but there are recognisable stages in the degeneration of a water body. In very general terms, when the water is around pH 6, snails, crustaceans, and molluscs—creatures which form shells—find it difficult to survive, as do some types of plant and animal plankton. As the pH moves below 6, salmon, trout and roach start to disappear; at pH 5.5 the whitefish and grayling begin to go, followed at pH 5.0 by perch and pike. The eels keep swimming around until pH 4.5, below which they also start to disappear, leaving the lake to certain insensitive insects (those with thick skins, air-breathers, etc.) and those animal and plant plankton which have not perished.[17]

There are parallels with the destruction of the forest ecosystems here:

- some species are particularly sensitive to acidification and disappear early in the pollution cycle;
- certain species are more resistant than others and continue living to a very late stage of acidification—indeed, certain plankton and insects proliferate and take over the new, degraded environment;
- some animals or plants die not because of their vulnerability to acidity but because they are dependent on other plants or animals for their survival, and these have perished;
- the end result of the fight against acidification is a 'dead' ecosystem, capable of maintaining a small number of species but in no way approaching the teeming, varied ecosystem that had existed previously.

Acidified lakes are not dead in the sense that 'no life exists at all'. There are no *fish* left, however. A 1973 survey of three acidified lakes in Sweden found 0, 1.5 and 1.5 perch per hectare of lakewater. The normal population of perch per hectare in non-acid lakes is several thousand.[18] In time, these few survivors would die also, leaving the lake devoid of any life-forms larger than insect-sized.

What do acidified lakes look like? Paradoxically, they are sometimes very beautiful. Sphagnum and other acid-loving mosses and algae invade the ecosystem and form a thick carpet along the lake bottom, trapping sediment and dust. The water, undisturbed by fish activity stirring up clouds of mud, becomes perfectly clear so that it is possible to see much deeper into an acid lake than into a normal one. Water lilies, existing by virtue of their roots embedded deep into the lake bottom, nestle comfortably on the surface. There is a quiet, picture-postcard quality about the scene. It is the peacefulness of death, however, rather than the calm of natural life.

Fish do not always die in acid water. If acidification occurs over a very long time they are capable, to a limited extent, of adapting to it—and in fact have done so in various parts of the world. Tolerance to acidity is something which can be evolved and inherited[19]—but time is essential. If the normal evolutionary process is short-circuited, then the fish cannot survive. Needless to say, the amount of acidity which we are dumping on our aquatic ecosystems, and the time-scale in which we are doing it, bears no relation to normal evolutionary processes.

Acidification can kill quickly or slowly depending on how it occurs. In a country like Norway, for example, there will be a large build-up of snow during the winter, containing a fair amount of acidity. When the spring snowmelt occurs, this is released into lakes and streams. The first 30 per cent of meltwater contains 50-80 per cent of the total amount of acids, metals and other substances in the snowpack: an enormous barrage of acidity which suddenly comes charging into the water system. The resulting pH reduction can range from 0.2 to 2.0 units (i.e. a 100 per cent increase in acidity) and may last from one day to two or three weeks.[20] This 'acid surge' hits the fish population at its most vulnerable—when the young are hatching—and can wipe out an entire new generation of fish. Acid surges occur either during the spring snowmelt, or in autumn when the first heavy rain after a long dry summer washes accumulated pollutants from the soil into the nearest waterbase.

Acidification can also kill slowly, by combining with certain forms of aluminium, which are washed into the water system from the acidified soil. The aluminium content of the water generally increases as the lakes and streams become more acid, and the combination of dissolved aluminium and water at a pH of around 5 is particularly lethal.[21] The effect on fish is to clog their gills with mucus, impede their respiratory systems, and lower the salt content of their blood. It has only recently been realised that the real cause of mass fish deaths in acid lakes is probably aluminium poisoning. This is linked to acidification because aluminium is not normally present in such large quantities; it has been carried into the watercourse by acid rain running through the soil. If there is a certain amount of calcium present in the water, this can offset the effects of acidity and help the fish to survive.

Of course, other members of the ecosystem will suffer either as a result of direct poisoning or from lack of food when their natural source disappears. In a moderately acidified lake, the younger fish die first leaving generations of older and larger fish with less competition for food. One might expect these older fish to become enormous as they no longer have to compete with thousands of young fish to find sustenance, but this does not happen. Brook trout reared in acidic water attain half the bodyweight of similar specimens reared in non-acid water. It seems that the energy cost of living in an acid environment is high and leaves less surplus energy for growth.[22]

However, the young fish, a very important food source, are no longer available. This affects birds such as herons and divers, and other water birds will experience the disappearance or reduction of important organisms from their normal diet. The consequences of this will not be immediately apparent due to the opportunistic feeding habits of many species. In the long term, however, effects could range from a decline in the population of certain birds to total local extinction. One bird, the pied fly-catcher, is already suffering fatalities as a result of acidification of its feeding grounds around certain streams and lakes in Sweden. Insect larvae living on the bottom of streams absorb aluminium from the acidified water which is transferred to the fly-catcher when it eats the adult insect, and accumulates in the bird's body. Soon, the female bird starts laying eggs with thin shells which are not capable of hatching into young birds. Eventually the parent dies[23] for reasons which are not fully understood but are, one presumes, connected with aluminium poisoning and calcium deficiency. Dr Erik Nyholm of the University of Lund in Sweden, studying the breeding of small songbirds by lakes in Lapland, found fewer eggs, less hatching success, and soft or missing shell material. He theorises that the birds were poisoned by aluminium from feeding on contaminated insects.[24]

Frogs and salamanders die out also at lower pH levels, because they experience breeding difficulties. Some species of plants and animals take the opportunity afforded by the absence of competitors to expand their population as much as possible: golden-eye ducks may proliferate on a lake surface because the lack of fish means an abundance of their favourite food—swimming insects like Hemiptera. So acid lakes can be beneficial to certain species.

To sum up:

- fish populations start to decrease at certain levels of acidity;
- the lethal character of acidification can be accelerated by the presence of certain substances such as aluminium in the water, or slowed down by the presence of calcium;
- the pH levels of some rivers and lakes do not stay the same all the year round, and can be altered dramatically by an 'acid surge';
- fish can learn to tolerate acid water to a limited extent, but only after a long evolutionary period—sudden changes in acidity will kill them;
- the consequences of rapid and continued acidification currently being experienced in water bodies throughout Europe, is the eradication of all fish in some ecosystems;
- the number of acidified lakes has multiplied since the early 1960s, and according to one report will double by the

year 1990 unless remedial action is taken;

● destruction of aquatic systems necessarily has a knock-on effect to non-aquatic species, especially fish-eating birds, but it is too early yet to quantify exactly what is happening.

So how widespread is this phenomenon? Is it just occurring in Scandinavia, or is it spreading to other countries? To answer this question, it is helpful to go back along the ecological chain and look at the distribution of acid rainfall.

Figure 2.1

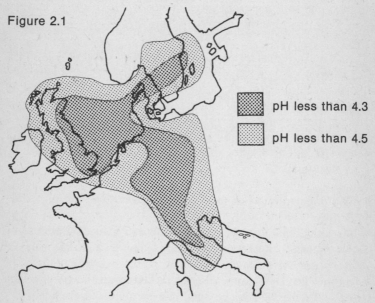

pH less than 4.3

pH less than 4.5

Figure 2.1[25] is based on OECD calculations, and shows the 1974 acidity of precipitation over parts of Europe and Scandinavia. The pH of 'normal' rain is about 5.6. This map shows a heavily acidified area of rainfall (pH 4.0 to 4.5, 10–80 times more acidic than usual) extending from the central part of Europe over Britain and the North Sea, and into southern Scandinavia. The rain that comes down in parts of Britain is in fact as acidic as that which has done enormous damage to swathes of Scandinavia and North America,[26] but because of the alkaline buffering qualities of

British soil, large-scale acidification has not yet been reported. In the southern areas of Sweden and Norway, however, where the soil is thin or poor, acidity has taken a strong hold. It is also spreading rapidly, affecting more and more water systems.

Dr Ivar Muniz is a Norwegian scientist who has been studying the effects of acid rain in Norway for some years. In 1982 he reported to the Stockholm Conference on the Environment:

> There is today a progressive acidification of freshwater in Norway, with no signs of levelling off. Data on changes in fish stocks in our southern counties show that more than half the fish populations are now lost. For the brown trout, more than 2,000 populations are lost or in the process of going extinct. The rate of disappearance has been particularly rapid from 1960 on, and if this trend persists, more than 80 per cent of the trout populations in the southernmost counties will vanish in this decade. The acidification is now clearly the most serious and extensive environmental problem in Norway ... Considering that fish is only one element in the aquatic system, the situation may well be characterised as an ecological disaster.[27]

In 1.3 million hectares (5,000 square miles) of Norway's southern lakes, the fish are practically extinct and inland stocks have been affected in nearly another 2 million hectares (7,500 square miles)[28] — a total area bigger than Belgium. In Sweden, 20 per cent of the lakes are affected, and fish in at least 4,000 of Canada's lakes have died as a result of acid-carrying pollution.[29] There are also large-scale aquatic deaths in certain parts of the USA, and increasing reports of acidified lakes in the UK.

How much does all this destruction cost? The problem is the same as for the acidification of the forests. Is an acid lake just a collection of dead fish, to be assessed according to their market value, or is it a symptom of ecological disaster? What price do you put on a dead lake?

The OECD, working only on the cost of fish losses, produced a figure for 1980 of $28 million a year for Scandinavia;[30] and EEC-sponsored research uses this figure to make its own estimate of $30 million (£20 million) for Europe. This latter report, however, adds an interesting aside: 'Estimates of loss of tourism value from lake acidification are extremely difficult to make and possibly should be ignored overall. The argument against taking loss of tourist expenditure into account is that tourists will spend their money elsewhere.'[31] The researchers have not attempted to work out the cost to the community of having large expanses of poisoned water in its midst. They consider trying to calculate the loss to the local economy from the lack of tourists but decide against this on the grounds that the money will be spent elsewhere. This is a strange proposition to put to a country which is losing revenue as a result of other countries' pollutant practices. It would be interesting to see the response to this argument by the shop-keepers of Brighton should a foreign tanker spill oil all over the beaches of southern England.

There are other ways of counting the cost of acidification of lakes and rivers:

Teigland ... found ... that in 1974 about 750,000 persons above the age of 15 were engaged in freshwater recreational fishing in Norway. On the average, each person spent 63 hours fishing during a fortnight each year ... A more general evaluation of welfare losses due to acidification of freshwater is made by Strand (1981). By assessment of the time used connected with this activity and by using data from a case study of one particular river, the value of freshwater recreational fishing in Norway is estimated at 700−1,000 million Norwegian kroner per year (US $110−160 million). [32]

If the fish in all these lakes are wiped out, then recreational facilities worth $110−160 million per year (£93−135 million, 1984 prices) disappear too: yet the OECD and the EEC-sponsored research have estimated the *total* cost to Scandinavia and Europe at $28 million and $30 million respectively. This is just one example of how acidification analysis can yield different costs, depending upon how broadly the pollution effects are considered.

Canada has started reacting vigorously to acid precipitation originating from the USA, as it *has* realised the economic implications. John Roberts, the Canadian Minister for the Environment, made a speech on 11 April 1983:

> The Canadian economy is dependent on our natural
> resource base. This natural resource base is threatened by
> acid rain. The gross economic activity generated by sports
> fishing in eastern Canada exceeds $1 billion a year.
> Tourism revenues are more than $10 billion and forestry
> produces more than $15 billion dollars a year. Altogether,
> we are talking about a substantial portion (about 8 per cent)
> of our gross national product ... at risk because of acid rain.
>
> A large proportion of Canadians live either within or
> very close to areas where acid rain is a real issue. A lot of
> Canadians therefore make a direct connection between
> themselves and the harm that acid rain can do.[33]

This 'direct connection' is an interesting though not illogical
phenomenon in the debate on acidification. Governments generally
adopt a sceptical posture towards the idea of acid-based pollution,
arguing that the effects are exaggerated, not enough is known about
the problem, and much more research is needed. This attitude provides a good defence against taking air pollution precautions, so
long as the people in the countries concerned believe that the debate
is about an abstract scientific matter which does not affect them
directly. The moment they start to suspect that their own environment is being threatened, then the political pressures generated
can make even the most intransigent administrations move very
quickly indeed. This happened in West Germany in the space of
three years (see page 79).

The UK may well be the next in line to experience this outpouring
of public concern. All the foundations for acidification are already
in place: widespread acid precipitation; large areas of coniferous
forest which, even if they do not perish themselves, attract acid
deposition and then route it into the nearest watercourse; and, most
important of all, evidence that certain of our lakes have increased
their acidity by a factor of 20 or more in the last twenty years.[34]
Parts of Scotland, Wales, and the Lake District have become progressively more acid, and some fish deaths have already been
reported.

Is this the beginning of large-scale acidification in the UK, or just a few isolated examples? A Swedish journalist by the name of Bo Landin, who has been specialising in acid precipitation for the last ten years, toured the UK recently to check up on the progress of acid destruction. He told a House of Commons meeting in July 1983:

> We saw exactly the same things we had observed in Scandinavia ten years ago. The fish disappear; they have the same symptoms; the same insects are gone; the same insects proliferate or survive. The changes are exactly the same as we saw in Scandinavia.[35]

The extent of acid damage in the UK is discussed on page 89 — but the point should be made here that as far as the UK is concerned, acidification of lakes and streams is not just a quarrel between foreigners. It is happening in the UK and has already cost at least one fish-farmer thousands of pounds. The Scandinavian and North American experience indicate that, once this pollution takes a grip, it gets progressively and rapidly worse. It could well be our turn next.

Materials, buildings and monuments

In 1978 UNESCO listed Cracow in Poland as one of 12 sites and monuments worthy of preservation in terms of the world's cultural heritage. It has 1,100 buildings which date from the eleventh century to 1830, and its 35-acre old town, complete with cobbled streets and the largest central market in Europe, is often said to be the finest surviving example of a medieval habitat.

The city has another claim to fame, however. The Cracow region, which covers 11 per cent of Poland's total area, emits 56 per cent of the country's pollutants and 48 per cent of all gases. In the drive for reconstruction after the second world war, the government rebuilt war-damaged cities such as Warsaw, Gdansk, and Posnan, but identified Cracow as a prime site for industrialisation. A workers' city and a steel complex, Nowa Huta, were built in the early 1950s, and later a mammoth aluminium foundry went up ten miles to the south-west. The smelter's capacity was originally projected at 15,000 tonnes per annum, but production steadily increased, reaching 53,000 tonnes in 1980. As output expanded, the plant was extended: old equipment was kept going after it should have been retired, no improvements were made in the production process, and there was a complete lack of anti-pollution facilities. The prevailing wind over the complex is westerly.

Cracow sits in a bowl surrounded by hills, and is one of the most polluted places in Europe. Its industries and heating plants emit some 150,000 tonnes of chemical waste into the air every year, which, because of the particular climatic conditions of the area, are held in a layer 80−100 metres above the ground. The old city of Cracow has up to 150 days of fog every year.

Large quantities of sulphur dioxide and hydrogen fluoride eat away at architectural detail in the ancient buildings, corrode stone and stucco work, and seriously weaken walls, roofs, and many metal elements. The pollution has eroded the faces of thirteenth-century stone statues and is dissolving the wood figures of the city's gothic cathedral so quickly that some experts have demanded that all important ornamentation be stripped off and stored in museums. Normal acids do not corrode gold, which in the laboratory is dissolved by a mixture of hydrochloric and nitric acids known as 'aqua regia', or royal water. In Cracow, air pollution is so strong that it has dissolved the gold roof of the cathedral chapel at Wawel Castle.

The twelve apostles no longer bless passers-by from their high stone plinths outside the sixteenth-century church of Saints Peter and Paul on Grodzska Street: the statues have been taken down and put in wire mesh cages in the courtyard, their features wiped away by acid, until the authorities decide what to do with them. 'We may put them into a museum,' said Dr Julian Fanfara, a research specialist with the Cracow historical reconstruction agency. 'If we were to somehow reconstruct them and put them back, in five years they would be damaged again.' Between 1975 and 1980, $3 million was spent on building renovation; 50 buildings have been reconstructed, and work on 60 others is under way.[36]

This information is not new−most of it comes from an article in the *New York Times* of 8 June 1980−and is not going to create any surprises in architectural circles. Corrosion of buildings is a problem that has been exercising the minds of government and local authorities for some time now, with sulphur dioxide once again being identified as the main culprit.

Strictly speaking, the erosion of some of the finest buildings and monuments in Europe is not primarily caused by acid rain, but by local emissions of pollutants. Dry deposition—the direct depositing of sulphur dioxide on to building surfaces, as explained on page 10—is the major cause. The argument over acid pollution often implies that acid rain and dry deposition are two completely different pollutants: but they are created from the same source—the burning of fossil fuels—and sometimes have interchangeable effects. An important difference between the two is their ability to travel: acid rain can transport itself hundreds of miles through the air, whereas dry deposition is usually a locally-created phenomenon, coming from a source which is only twenty or thirty miles away.

The destructive powers of acid pollution are awesome. There are buildings which had weathered the elements for two thousand years before the introduction of local industrialisation; but once smelters or power stations appeared in the vicinity, the resulting corrosion produced more deterioration in twenty years than had happened in the previous twenty hundred. The pollution erodes metal structures, certain kinds of stone, leather, paper, and textiles. It can rub the face off a statue, strip paint off metal surfaces, take hundreds of pounds off the value of a car. According to a story in *New Scientist*,[37] ICI used to give its higher-paid employees shiny new cars—but in their chemical plants in Runcorn, all the parking spaces were taken up by old Morris Minors and second-hand Cortinas. The car-owners had noticed that the atmosphere was corrosive, mainly from the waste acid produced by the sulphuric acid plant which was vented to the outside air. Cellulose paint is not proof against acid attack and it was reckoned in the factory that a car's shine would be ruined in two years. So people left their new automobiles at home and drove to work in old bangers. (The problem is solved now—ICI scrubs its wastes clean.)

All European countries are paying enormous amounts of money to counter the effects of acid-based atmospheric pollution—in many cases, without realising it. As an example, paint deteriorates very quickly in a polluted atmosphere: galvanised steel needs to be painted every four years in sulphur-loaded industrialised areas, but only every twenty years in a rural atmosphere. The corrosion rate of zinc is six times higher in polluted air than when the air is clean.[38] Carbon steel, nickel, and copper corrode more quickly in similar circumstances.[39] Electrical devices, particularly those containing copper, silver, and gold contacts, show increased failure in polluted zones. Cars in towns have a shorter life than in the country. More expensive protective measures are necessary for steel structures in urban sites.[40]

All this deterioration and corrosion has to be paid for, but the problem from the environmentalists' point of view is that the important part played by fossil-fuel pollution in this process is rarely acknowledged. If people knew that they had to paint their windows two years early because of sulphurous atmospheric pollution, then local pressure towards desulphurisation would be a lot more tangible.

No one has yet compiled a world-wide analysis of the extent of material damage caused by acidification, though work is in progress on a European list. The examples below are intended to be illustrative rather than comprehensive.

London
In London every large historic building is suffering from acid-based damage. In some places more than an inch of the Portland stone used to build St Paul's Cathedral has been eaten away. The north-west tower was recently restored at a cost of £300,000.[41]

The Netherlands
In the Netherlands, the pictures and ornaments on the exterior of the 458-year-old Sint-Jans-Kathedrale, 's-Hertogenbosch, is said to be 'melting like toffee'. The building is under permanent restoration. Damage to historic buildings here is estimated at 30 million guilders (£6,750,000) per year.[42]

West Germany
In West Germany, the historical mining museum in Dortmund loses 4 per cent by weight of its structure every year. Damage to Cologne Cathedral alone costs an annual £1,500,000.[43]

Italy

In Rome, Michelangelo's statue of Marcus Aurelius—a symbol of the city—has had to be removed because of the corrosion of the metal.[44] The Colosseum has been 'seriously damaged'.[45] The Arch of Titus has been affected. All through Italy, accelerated deterioration of stone materials is being widely reported: in Venice, the Basilica of San Marco and other buildings are being corroded by sulphur-bearing pollutants from the industrial north of the city. In Florence, the fall of fragments from ancient buildings and monuments is creating serious problems for the authorities, even to the extent of endangering public safety in the street.[46]

New York

In New York, Cleopatra's Needle, an ancient obelisk obtained from the Khedive of Egypt in the 1860s, still has a clear inscription on its east face but the writing on the west, which faces the prevailing winds, has been erased by the chemicals in the air. Ninety years in New York has done more damage to the stone than 3,500 years in Egypt.[47] The Statue of Liberty is now so heavily corroded that an emergency rescue operation has had to be undertaken. During the next two years, restoration work estimated at $30 million (£20 million) will be carried out. This is approximately six times as much as the entire statue cost to erect in 1886.[48]

Sweden

The cast-iron spire of Stockholm's Riddarholm Church suffered so much corrosion that it had to be replaced in the 1960s. The church's Royal Burial Chapel is now rapidly deteriorating. The erosion of limestone in 100 medieval churches on the Swedish island of Gotland in the Baltic became so acute that the authorities forced local industries to use only low-sulphur fuels.[49]

India

The Taj Mahal is dusted with 13 tonnes of industrial emissions every day, from two power stations and from local foundries. Restoration experts have warned that corrosive deterioration has accelerated and that the monument may eventually disintegrate. At the moment, its marble is discolouring, the sandstone is flaking, and semi-precious stones set into the walls are blistering.[50]

Athens

The Acropolis in Athens is being eaten away by industrial smogs.

Corrosion is so bad that the four caryatids (female figures used as pillars) from the Erechtheion, the second largest temple in the area, have been removed to a museum to be stored in conditioned nitrogen. The Cecrops statue has been similarly taken out of harm's way. Parts of the Parthenon are to be dismantled and rebuilt around a skeleton of rust-free titanium alloy. Buses are now forbidden on the Acropolis hill; the whole of Athens has been prohibited from using high-sulphur fuel oil; a zone has been created around the most threatened buildings, in which solar energy or electricity are to be the sole sources for central heating for private houses; in future, only electric cars will be allowed in the area.[51]

The drastic measures taken in Athens reflect the irreplaceable nature of the monuments which are being worn away. All buildings in polluted areas are corroded by acidic pollution, but the effect of this depends on the character and intricacy of the structures that are being damaged. The erosion of half an inch of masonry from the outside of a solid suburban town hall is probably not going to cause undue concern to conservationists; on the other hand, finely-featured sculptures and statues cannot afford this amount of material destruction.

It should be pointed out here that sulphur compounds are not the only eroding pollutants: ozone, hydrocarbons, carbon monoxide, and carbon dioxide (to name but a few) have deleterious effects on certain materials. Also ordinary rain, wind, and atmospheric humidity are major participants in the eroding process and some people argue that the effects of SO_2 are insignificant in comparison with these natural causes.

This latter point was made at the April 1983 meeting of the European Parliament Environment Committee, and attracted the following response from a participant: 'Atmospheric pollution is the main cause of the destruction of works of art. Humidity, wind and rain have always existed, but now ancient buildings are suddenly crumbling within a decade.'[52] This sentiment was echoed in the testimony of a Greek specialist on acid corrosion when he gave a paper on the subject to the Karlsruhe EEC symposium on acid rain in the following September. He described the methods used to identify when the deterioration of statues in Athens began, and said:

From the above-mentioned observations it was concluded that the severe deterioration began in the last 20−25 years, coinciding with the beginning of the intense industrialisation of the Athens area. Thus the pollution caused, during this period, more deterioration than the exposure of the monuments for 2,400 years without it.[53]

Both statements are linked to the recent destruction of forests and lakes. Humidity, wind, and rain have always existed, with foreseeable consequences. What we are facing now is a completely new phenomenon which has made itself apparent over the last two or three decades.

The other question — how much sulphur dioxide is responsible for this destruction of buildings and statues, given that other pollutants exist in urban atmospheres — has been addressed in the Netherlands. In 1980, a study estimated the economic damage to materials due to SO_2 by examing a representative group of recently renovated buildings. It was concluded that 30−50 per cent of the renovation costs to the exterior of these monuments could be regarded as caused by sulphur dioxide.[54]

The OECD has carried out some work on the economic costs of sulphurous pollution on galvanised steel and painted steel, conceding that many other materials are also affected by atmospheric pollution but arguing that the cost of these could not be estimated, either because they have minor economic value or because the detailed relationship between deterioration and SO_2 has not been precisely established. Notwithstanding the restrictions of its survey, it concluded that a 37 per cent reduction of sulphur emissions would result in a saving of $963 million (1979 prices; about £900 million in 1984 prices) in the eleven OECD countries in Europe.[55]

This is not the true cost of material corrosion as the analysis covered only two substances. Sandstone and limestone were excluded from the calculations which meant that damage to buildings was not estimated. Historical artefacts were also left out on the grounds that cultural loss could not be expressed economically. The report says in mitigation:

Some attempts have been made to estimate the costs of restoring affected objects and buildings: figures of several hundred thousand dollars have been suggested even for single small objects. Estimation of the total economic losses at present seems to be difficult because no regional inventories of the materials used have been taken. Moreover, any restoration implies a decrease in originality of the objects and therefore a decrease in cultural value, a loss that cannot be estimated in economic terms.[56]

Once again the cost-benefit analysts find themselves incapable of estimating even the approximate cost of acidic destruction. The OECD acknowledges that it is likely to be enormous, however, and even spends some time preparing the ground: 'Sandstone and limestone are severely affected by SO_2,' the report tells us and then goes on to say that the two substances:

were, in the past, often used as building materials, especially in large constructions like churches and government buildings. They were also used for decorative purposes on entire buildings, portals, façades, and for ornamentation. Such decoration is often of cultural interest and, as such, has a substantial cultural value.[57]

So it would seem that there is a vast cultural loss at stake representing an extremely large sum of money. Having set the scene, however, the OECD declines to estimate the amount that is involved: 'the cost estimates presented here probably underestimate the true cost.'[58]

How much *is* the final price? The United Nations Economic Commission for Europe reviewed the literature and stated:

A number of national and international studies have shown that economic losses due to atmospheric corrosion caused by sulphur compounds are considerable. In these studies such losses have been estimated to range between $2 and $10 per capita per annum, or between 0.10 and 0.23 per cent of Gross National Product.[59]

These figures would give a cost of materials damage in the UK of from $111,352,000 to $556,760,000 per year (£83 million to £415 million, 1984 prices); and the European total, assuming a population of 650 million, of $1,300,000,000 to $6,500,000,000 per year (£1 billion to £5 billion, 1984 prices) — and this, it should be remembered, is just the cost of damage to materials and ignores all other losses incurred in lake and forestry degradation, not to mention other costs to be discussed in the next section.

Costs of corrosion increase as research considers other areas. On 28 February 1984, the *Guardian* reported a new discovery:

> *Geneva*: Acid rain is having a 'disastrous effect' on Europe's stained glass treasures and could destroy them entirely, a UN study says.
>
> More than 100,000 stained glass objects, some of them more than 1,000 years old, are gravely threatened, it warns.
>
> ... The study by the UN Economic Commission for Europe shows ... that stained glass objects were generally in good condition up to the turn of the century. But it says that, in the last 30 years, 'the deterioration process has apparently accelerated to the extent that a total loss is expected within the next few decades, if no remedial action is taken ...
>
> It also warns that although many precious objects made from these materials are in museums and protected from corrosion, damage from air pollution has been 'far from negligible'.

Working out the price of pollution is an extremely difficult task; at least the OECD report admits that its calculations were probably underestimates. Perhaps the situation is most adequately summed up by Herr Muntingh, who compiled the background Information Note for the 1983 European Parliament hearing on acid rain: 'The problem of acid rain, in terms of damage to monuments, buildings, libraries, etc. is worth billions of guilders — or billions of any other currency you care to mention.'[60]

Health, crops and water

Health
Thorpe's *Dictionary of Applied Chemistry* describes sulphur dioxide as:

an irritant, irrespirable gas, relatively non-toxic in comparison with carbon monoxide or oxides of nitrogen. Very high concentrations, when the victim cannot escape, may cause death from respiratory spasms and asphyxia. *A mild degree of sulphur dioxide poisoning* produces headache, anoxia, spasmodic cough, sneezing, hemoptysis, bronchitis, constricting of chest, gastrointestinal disorders, conjunctivitis, smarting of eyes, lacrimation, and anaemia. Ulceration of the mucous membrane may also result. [emphasis added]

The World Health Organisation adds that SO_2 is a highly soluble gas which, when inhaled, disperses readily in the bloodstream. Breathing functions are affected at relatively high levels of atmospheric pollution, but sensory changes and altered brain responses can occur before these respiratory effects take place.[61]

The OECD notes that there is 'a causal link between atmospheric pollution and human health', though it believes that another product of the fossil-fuel chain, sulphate (SO_4) is the main active agent producing adverse effects. SO_4 represents a small but significant fraction of suspended particulates (i.e. dust, etc.) in the air. The organisation conducts its analyses on this assumption, while conceding also that 'there is considerable evidence linking SO_2 and total suspended particulates with health effects'.[62]

The exposure of communities to sulphur dioxide and smoke has been investigated in several scientific inquiries. The most clearly defined occurrences have been when people died as a result of poisonous smogs (Meuse Valley, Belgium, 1934; Donora, Pennsylvania, 1948; London, 1952),[63] and these episodes led to a prolonged series of air pollution studies.

Mainly as a result of the findings of these researches, air quality in European cities has improved since the 1950s. Tall chimneys pipe the pollution out of the immediate vicinity with a resultant improvement in atmospheric conditions around the emitting source. There are fewer deaths now, but there are still fatalities arising from air pollution and a significant amount of illness too.

As the OECD points out, the normal reaction to pollution is not one of life-or-death. There are intermediate stages of physical and psychological stress and illness before a person succumbs. Using an analysis correlating annual average levels of sulphate particles in the air to known occurrences of illness, the OECD tried to work out the cost of sulphur-based pollution on human health. This is difficult, as it involves calculating the financial consequences of hospitalisation, out-patient care, loss of income, loss of productivity, sickness cash benefits, and so on—at the same time taking into account other factors such as co-existing pollutants, social background of the people affected, age, occupation, medical history, etc.[64]

Having given all these caveats, the statisticians worked out that in 1974, if the UK had had 37 per cent less sulphur emitted into its atmosphere, it would have saved between $60,420,000 and $1,510,400,000 in health care costs and lost income—and that an average person's life expectancy would have been 11—275 days longer.[65] The figures show such a wide range because of the complexity of the calculations, but they are an interesting parenthesis to the other economic and social losses arising from the use of fossil fuels, as discussed in previous chapters.

There is still room for improvement in the air quality of the UK especially in London. The World Health Organisation laid down guidelines for exposure limits consistent with the protection of human health. Over 24 hours, the mean should be 100—150 micrograms of sulphur dioxide per cubic metre of air, with the same quantity of smoke; and over a year, the mean should be 40—60 micrograms per cubic metre of both. Levels above this are regarded as potentially harmful.[66]

In 1976—77 *each one* of the 33 authorities in Greater London recorded an annual mean sulphur concentration in excess of these levels; 1,000 square kilometres (385 square miles) of London was estimated to be over the limit, with 100 square kilometres (38 square miles) exceeding twice this value.[67]

The trend for atmospheric pollution since then has been improving, but national governmental attitudes towards the health of their citizens are still worrying. In 1980, the EEC recommended that member states should conform to the World Health Organisation guidelines, but in the meantime established 'limit values', not as strict as the WHO values, which should be regarded as mandatory.

The intention is that all EEC states should aim for the WHO guidelines, but in any event should conform to the EEC limits.[68]

Largely as a result of changing energy patterns (mainly the move away from domestic coal) and probably also because of the recession, SO_2 emissions in London dropped to just below the EEC limit values in 1980−81.[69] The Greater London Council took a cautious line towards this drop, arguing firstly that the monitoring sites (National Survey of Air Pollution sites, run on behalf of the Department of the Environment) were not generally situated in the best places to give a correct pollution picture − for example, more sites should be placed near main roads to pick up the increasing emissions from the use of diesel vehicles. Secondly they argued that the drop in SO_2 pollution was not the result of government policy, but was based on a change of fuel preference and an economic recession. The corollary of this observation is that if energy uses change again in London, or the economy takes an upturn, there is nothing to stop an increasing level of air pollution with concomitant harmful effects to health. A GLC/Department of Industry working party recommended that, to prevent this happening, industrial fuel oil used in central London should not in future be allowed to contain more than 1 per cent of sulphur. This has been done successfully in the City of London in 1972.[70]

The Department of the Environment (DoE) looked at the dropping pollution rate and decided that such a restriction was unnecessary. It also decided in 1983 to cut the number of monitoring sites in the UK by 90 per cent. The GLC had argued that if these sites were in a different spread of locations in London, they would have given higher pollution readings and thus not presented such an optimistic assessment of the capital's air quality. The response of the DoE was that the readings they already had were within EEC limits and the trend in pollution was falling, so the monitoring sites were redundant.

Meanwhile, the GLC's own monitoring sites showed that parts of London in 1982 and 1983 were still exposed to air pollution above the WHO recommendations − for example, York Road in Lambeth and East India Dock Road in Tower Hamlets.[71] A spokesperson in the GLC said, 'In the UK we have traditionally had a reactive policy towards air pollution rather than a preventive one. The GLC wanted to take measures to ensure that air pollution did not return to London, and that if it did, it would be monitored quickly.'[72] The Department of the Environment, it appears, did not share that concern. The sulphur dioxide concentrations in central London remain the highest in the UK.[73]

Internationally, there is still much anxiety over the effects of sulphurous air pollution. According to an American Friends of the Earth publication (*Alternatives*, Winter 1983) a 1982 US Congressional Report indicated that 51,000 people died in North America in 1980 from illnesses caused by airborne sulphur compounds. The victims are people who already suffer from respiratory disease, says the report's author, Robert Friedman. In Canada, researchers in Ontario have been told to find out what effect the thousands of tonnes of sulphate in the American atmosphere has on human health. The environment ministry of Ontario does not want to publicise the effects, according to Friends of the Earth, 'because it doesn't want to cause alarm unduly'.

In Japan there is a long-standing awareness of the dangers of sulphur oxide poisoning. The Pollution Related Health Damage Compensation Law was passed in 1973. Sufferers meeting the provisions of the law are guaranteed compensation which the polluting party has to provide. People who suffer from asthma and suchlike, caused by pollution, can seek redress. Compensation awarded as a result of sulphur oxides discharged in 1977 amounted to approximately 41 billion yen (over £120 million).[74]

Fogs are again being examined with regard to their damaging effects on health, especially acid smogs on the USA's west coast. An American Heart Association meeting was told in 1983 that hot, smoggy days could affect the health of the spectators at the 1984 Olympic Games in Los Angeles. Those with heart diseases would be particularly affected according to Dr Peter Snell, a University of Texas medical researcher and former athlete.[75] In August 1983, the American Lung Association issued a call for the reduction of sulphur oxide emissions:

A number of epidemiological studies show clear evidence of increased sickness and mortality in the population at large caused by elevated concentrations of sulphur dioxide and particulate matter in the ambient air. On a national basis, at high pollution levels, fractional increases in sulphur dioxide and sulphate concentrations may well take an increasing toll in illness and death, particularly among the elderly, persons with respiratory and cardiovascular disease, and other susceptible individuals.[76]

Californian research is concentrating on acid fogs, especially in relation to its effects on human lungs. Michael R. Hoffman, an environmental engineer, has found that fog in the Los Angeles area is often 100 times more acidic than the rain. He has developed his theories to predict the likely acidity of the 1952 London smog, which he estimates to have had a pH of between 1.4 and 1.9 — extremely acid and, he believes, possibly an important cause of the 4,000 deaths during and after the smog occurrence.[77] This kind of present-day pollution is not limited to North America. On 2 January 1984, Athens suffered yet another of its crippling smogs and as a countermeasure, central heating was switched off in banks, schools, and public buildings, and traffic was banned from a central zone in the city. Athenians with lung problems were warned to stay home until the pollution cleared.[78]

The World Health Organisation has warned of the suspected link between air pollution and lung cancer,[79] and this is being actively researched on the east coast of the USA. Seven counties along the coast from Baltimore to New Orleans are among the ten urban counties nation-wide with the highest lung cancer deaths for white males. Atmospheric chemist John Winchester and his colleagues at Florida State University have noticed that lung cancer fatalities among white males in this area nearly tripled between 1950 and 1975. 'Acid air', formed when sulphur dioxide encounters water vapour in the air and turns into a sulphuric acid haze, is being blamed. The researchers estimated the concentrations of acid air over three states and found that sulphuric acid particles increase in concentration to Jacksonville, Florida, then decrease with an influx of clean marine air. Lung cancer mortality rates in the area follow a similar pattern. Winchester says: 'We found that if you look along the east coast from South Carolina to Florida you notice a gradual increase in the lung cancer mortality as you move southward to about Jacksonville, and then it tapers off again.' The working hypothesis at the moment is that the acid air works synergistically (see page 47) with cigarette smoking.[80]

Acid haze or acid air, representing a system of acidic transference to the ground which is neither wet nor dry, has been dubbed 'occult' deposition. The phrase, which has come into use quite recently, includes wind-blown cloud and fog droplets containing concentrated acidity, and represents another threat caused by unfiltered fossil-fuel products.

There are two more items of possible interest. First, West German scientists are investigating the likelihood of air pollution being a contributory factor towards Sudden Infant Death Syndrome ('cot deaths'): out of 639 babies autopsied in the Monchengladbach, Cologne, and Aachen areas after deaths for no apparent reason, 80 per cent had some kind of air-passage infection, especially inflammation of nose or throat.[81] Second, the Soviet Academy of Science is concerned that pollution is causing health problems in the Kuznetsk basin of southern Siberia, an area dotted with coal and chemical plants. 'It is turning out rubbish, driving people away and causing an accumulation of genetic defects,' according to Professor Vlail Kaznacheyev, head of the Institute of Clinical and Experimental Science at the Academy.[82]

Crops

Damage is done to crops by a mixture of pollutants—sulphur dioxide, oxides of nitrogen, and ozone. Ozone (O_3) occurs naturally in the atmosphere, but is concentrated mainly in a belt 7−16 kilometres (4.5−10 miles) above the Earth's surface, where it shields the planet's surface from ultra-violet radiation. Some O_3 is transferred to ground level by atmospheric turbulence, although natural concentrations at the surface are generally low. During summer, however, emissions of nitrogen oxides and hydrocarbons (a collective term covering many different molecular mixtures of hydrogen and carbon atoms) from motor vehicles and other urban sources may take part in chemical reactions in combination with sunlight in the air, producing ozone at ground level. Nitrogen oxides are produced in large quantities by fossil-fuel combustion, so this ozone production can once again be linked to the acidification debate—though the precise details of ozone production are still the subject of discussion.

The pollutants can work singly, in pairs, or all three together; current scientific thinking is that when two of these substances team up, their combined power is much greater than that of each one acting separately. This is called the synergistic effect, and can be expressed symbolically as $1 + 1 = 3$. The synergistic combination of sulphur dioxide and smoke has a harmful effect on human health; that of SO_2 and NO_x possibly damages trees, and certainly harms certain crops. Ozone also, in combination with the other two pollutants, seems to damage vegetation, though it can also produce results working on its own. It would appear that some of the damage done to the German forests is ozone-linked.

Research with plants in fumigation chambers has shown that certain vegetative growth can be reduced by up to 35 per cent in atmospheres treated with mixtures of SO_2 and NO_x.[83] The implications of this are disturbing for farmers and agriculturalists: if this air pollution is reducing plant growth and yield, the economic consequences may be considerable. A 1982 report has indicated that those crops which are quite sensitive to SO_2 pollution include barley, oats, wheat, beans, beetroot, brussel sprouts, carrots, leeks, peas, spinach, tomatoes, apples, pears, and red currants; and those which are definitely sensitive include rye, lettuce, parsnip, sweet peppers, radish, gooseberry, hops, and rhubarb.[84]

A note of caution should be sounded here, though, as modern farming practice injects a good deal of acidity into agricultural soil. Ammonium-based fertilisers, for example, are very popular and are commonly used. In Sweden these substances contribute about 20 per cent of the acidification of Swedish ploughland, as compared to 10 per cent by acidic deposition.[85] On the other hand, farmers also commonly spread lime on their soil—this has been happening in the UK for years—and they can calculate and balance their application of fertiliser and lime. Atmospheric deposition complicates the equation by injecting acidity without the farmers' knowledge.

On the perennial question of how much it all costs, it is still rather early to give a definite answer—partly because so little research in the field has been done in the UK, and also because, as in most other aspects of acidic pollution, the figure are continually being revised upwards. The OECD's calculations estimated that if the 1976 sulphur output had been reduced by 37 per cent the UK would have saved $55 million in crop losses for that year.[86]

In March 1983, a Scottish newspaper[87] carried an off-the-record briefing from 'agricultural scientists' that crop damage was costing the UK £25 million a year. In September that year the same source disclosed an estimate that the figures was nearer £100 million per annum. Neither estimate was officially confirmed.

Then in January 1984, Dr Michael Unsworth of the Institute of Terrestrial Ecology – the leading UK body researching into acidification – said: 'I estimate that sulphur pollution is costing UK farmers £200 million a year in crop losses' (*Farmer's Weekly*). Dr Unsworth had been responsible for co-ordinating a series of trials, funded by the Ministry of Agriculture, investigating the effects of sulphur pollution on winter wheat and winter barley. 'Even comparatively low concentrations of sulphur dioxide have been found to interfere with photosynthesis,' said Dr Unsworth, and as a result 'both growth and yield are affected'. It has also been found that sulphur dioxide can increase the susceptibility of crops to frost damage, disease, and pest attack. 'Yield losses are likely to be between 5 and 10 per cent in the worst-hit areas. If they had been any higher than this we would have picked them up by now. The problem is that the effects of sulphur pollution are generally subclinical and generally produce no obvious symptoms.'[88]

Dr Unsworth's assessment agrees with recent US estimates. In June 1983, the White House Science Office released a report from a panel of nine scientists appointed in 1982 by Presidential Science Advisor, George A. Keyworth, to review the state of knowledge about acid rain. The Acid Rain Peer Review Panel looked at the effects of air pollution on the growth of crops and concluded that, although evidence was scanty, 'an estimate for ozone damage to agriculture in the United States is 5 per cent of the cash value. We anticipate that the overall effect of acid precipitaiton on crops could be comparably significant.'[89]

The United Nations Economic Commission for Europe has warned that growth of rye grass and barley decreases by 25 – 40 per cent in sulphurous atmospheres. The USA Office of Technology

Assessment has estimated that if ozone were reduced to its 'natural' levels, the value of increased peanut, soya bean, wheat and corn yields in the USA would be $3.1 billion per annum.[90] In the vast agricultural areas of the San Joaquin Valley, California, production of some crops has fallen by as much as 20 per cent and pollution, including acid deposition and ozone, is being blamed.[91] No one has yet attempted to calculate the full cost of air pollution on crops: the indications are that it is very large indeed.

Water

Human health has been adversely affected in Sweden by acidification of water supplies, mainly as a result of pollution from acidified groundwater. Acidity gets into private wells—more than one million Swedes use water drawn from a well of their own, and half of these live in areas sensitive to acidification—and the acidity can then affect copper piping, especially if the water stands in the pipes overnight. The Swedes have observed cases of diarrhoea in infants which have been associated with the presence of high amounts of copper in water. A Danish-based magazine called *Taraxacum* carried this report in 1982:

Anna Svenson lives together with her family in Bohuslan, a county just north of Gothenburg on the Swedish west coast. Since last summer they have been unable to use the water from their own well as drinking water.

'It was three years ago that we first noticed a change in the colour of the tapwater. It gradually turned more and more green. A little later at that time my two-year-old son, Peter, got ill with the stomach. He also got a very long-lasting diarrhoea. Later I also got a bad stomach.

'We brought samples of our tapwater to the health department of our municipality. They found that the pH-value was low, and that the content of copper was very high, about 5 milligrams per litre. [WHO's recommended highest value is 0.05 milligrams per litre of daily input.] They told us that the acid water was corroding the water pipes, which are made of copper. And they said it was most probably the high content of copper that had made me and my children ill.[92]

The acid well-water dissolves copper pipes used in most homes, leaks through cracks in the pipes and nourishes a fungus that grows in the wooden houses, causing expensive structural damage. Other increasing concentrations of metals in drinking water have been observed. Aluminium in water has caused fatalities in kidney dialysis treatment in Sweden, and the level of cadmium in certain supplies has increased by a factor of 100. The presence of copper in water sometimes has a bizarre side-effect: people with blonde hair have found that it may turn green after a shampoo.

Acidity of drinking water has not been identified as a problem in the UK as the water boards automatically treat our water to make it alkaline, if this appears to be necessary. There are still thousands of private wells in existence, however, and these will need to be checked to ensure that they do not become poisoned by mobilisation of metals. Medical opinion in Britain is just beginning to be concerned about the effects of increasing acidity in our drinking water. As *The Lancet* stated on 24 March 1984:

> The harmful effects of acid rain on aquatic ecosystems are now recognised: the potential effects on man, through his drinking-water, are yet to be defined. The leaching action of acid rain could, if allowed to continue and increase, substantially alter the toxic metal content of both surface and ground waters. Aluminium poisoning is recognised in patients with impaired renal function ... The toxic effects of the other leached-out metals are perhaps more clearly defined and better recognised as health hazards ... Clearly, the ravages of acid rain warrant urgent scientific and political activity.

3. Acid politics

Pollution, politics and pressure

Every country in Europe is emitting acidic pollution into the air, and each country is also suffering the effects of acid deposition. Some places, because of wind currents and the geological make-up of the land, have already reached the acid threshold point and are experiencing environmental degradation on an obvious and alarming scale. Other areas are believed to be relatively unaffected by the pollution; and this belief, unfounded though it may be in reality, has coloured national attitudes towards cleaning up industrial emissions.

Governments have divided their acid rain research into three areas: firstly, environmental and zoological enquiry, documenting the effects and growth of acid deposition; secondly, engineering research, which tries to develop ways of cleaning the pollutants from smoke; and thirdly, economic analysis, balancing the damages reported against the current costs of correcting the problem. The rationale is that the cost of cleaning smoke is not yet justified by the extent of the damage: there is time to wait until the cleaning technology gets cheaper, or the damages spread and change the scope of the problem, or both.

There are two flaws in this rationale. The first is that, in the equation, there is only one hard figure which can be stated with some certainty: the costs of cleaning smoke. Cost analyses of damage caused by air pollution are much vaguer and more difficult to quantify. Soft estimates of effects on lakes, forests, health and materials do not carry much weight when political decisions are being made, no matter how often the analyst points out that the figures considerably understate the extent of the problem.

Secondly, there is self-delusion in the way in which the rationale is presented. Governments are not trying to decide *whether* to pay the costs of acidic pollution; they are continuing an already established process of deciding *who* should bear these costs. As the head of the OECD's environment directorate put it at a conference in 1982:

> Acid rain is a good example of a problem that is fairly well understood and for which the technology of control, in the sense of a reduction of emissions, exists. The main issue is *not really* the balance between the costs and benefits of control, although it is often so represented. The *principal* source of controversy is *who* should bear the environmental costs associated with coal burning, and *how* and *when* those costs should be borne.
>
> We know that these costs may be borne by society at large: that is, by people, property and ecosystems exposed to the acid precipitation. That is generally the case now and they take the form of damage to human health, to property, and to deteriorated forest and lake ecosystems. Alternatively, the costs may be borne by the industry via investment in the technology available to reduce emissions at source. This is increasingly the situation today in many member countries and in this case a good part of these costs will be passed on to the consumers of electricity, or of steel, or of other goods. There is a third alternative which is not uncommon. The costs may also be borne by the taxpayer subsidising a part of the research and development underlying the technology and, perhaps, even subsidising its installation, in whole or in part.
>
> The point is that *the costs will be borne: they cannot be avoided*.[1] [Original emphasis]

The major culprits

There were about 65 million tonnes of sulphur dioxide emitted over Europe and Asiatic Russia in 1982: 21 million from the West, 40 million from the East, with 4 million from origins unknown.[2]

There are 26 large or medium-sized countries in the region (excluding places like Malta or Gibraltar) and of these, 11 nations produced almost 85% of the area's airborne sulphur dioxide. The polluters are Czechoslovakia (which pumped out 3.3 million tonnes of SO_2), the Federal Republic of Germany (3.5), France (2.9), the German Democratic Republic (4.0), Hungary (1.7), Italy (3), Rumania (2), Spain (2), the UK (4.2), Poland (2.5) and the USSR (25.5).

All these areas are contributing to the European and world-wide build-up of SO_2, and all are guilty of grossly polluting the atmosphere, using the world's airways as a cheap method for dumping their industrial effluent. In terms of the spoliation of Europe's ecosystems, however, some are more guilty than others. The general wind trend across Northern Europe is south-westerly,[3] so those people at the western edges of the continent are more likely to escape the deposition of SO_2 and NO_x from other countries, and less inclined to seek a clean-up of industrial emissions, believing themselves on balance to be benefiting from the current international exchange of airborne pollution.

Working out who dumps what on whom is one of the more challenging tasks of research into acidification. Switches of wind direction can change the path of a pollution plume; the amount of fuel a country uses can depend on the weather; the quantity of sulphur produced in the air depends on the sulphur content of local coal or oil. Two researchers into environmental pollution, Nicolas Highton and Michael Chadwick, estimated the SO_2 budgets of all the European countries for 1982, using a mathematical model.[4] Their calculations were done in 1982, which is rather more difficult than doing the work retrospectively. As a rough check, however, it is worth noting that their estimate for the UK, 4.2 million tonnes, is very similar to the Department of Environment estimate, worked out in 1983, of 4.04 million. Agreement to within 5 per cent on a calculation such as this shows a creditable accuracy of the original calculations, especially as the Department of Environment figure is acknowledged to be underestimated by two or three percentage points, because of the non-inclusion of certain industrial emitters.[5] The emissions and pattern of depositions in the 1982 analysis are generally within the range of statistics produced by the UNECE's monitoring programme (the European Monitoring and Evaluation Programme, see page 119).

Highton and Chadwick's chart, Table 3.1, shows that the USSR and Eastern European countries contribute a disproportionate amount of SO_2 to the atmosphere of Europe—about 61 per cent of the total. This is an inflated figure, however, as the 25.5 million tonnes of SO_2 that the USSR produces is the total for the whole country, not just from that part of it which is in Europe. Because of the USSR's position, 22.5 million tonnes out of the 25.5 million it produces, remain in Russia; and it is in fact one of the few large industrial states that actually imports more SO_2 than it sends out: 3 million tonnes are exported, 5 million are received from elsewhere.

The rest of the Eastern European countries are similarly disadvantaged geographically. On the other hand they produce, per capita, much more pollution than the Western nations. The situation has been described by Dr Krystian Seibert, Cracow's chief architect and planning officer, who told the *New York Times*:

> The capitalistic lands came to a high standard of development because of neglect of human and environmental rights. When they got to a certain level, they said: 'Now we are rich, boys. Let's wash ourselves off and become elegant.'
>
> We came late to industrialisation. We are trying to catch up and do it all at once. Some things are unavoidable. Fast industrialisation means devastation. But we don't want to destroy our old town.[6]

Much of the available fuel in Eastern Europe is brown coal, which has a higher sulphur content than the hard coal used in the West. Current per capita SO_2 production in the East is generally larger than, sometimes twice as much as, that of the West European countries. If there is to be a desulphurisation programme, it is essential that Czechoslovakia, Poland, Rumania, Hungary, the GDR, and the USSR take part in it. These countries contribute to the sulphur pollution of Scandinavia (30 per cent of the SO_2 in Sweden comes from these areas, 40 per cent in Finland and 23 per cent in Norway) and to West Germany (18 per cent) and

Table 3.1 Current emissions and depositions of SO$_2$ by country. (Thousands of tonnes per annum)

Receivers	Austria	Belgium	Bulgaria	Czechoslovakia	Denmark	FRG	Finland	France	GDR	Greece*	Hungary	Ireland*	Italy	Luxembourg*	Netherlands	Norway	Poland	Portugal*	Rumania	Spain	Sweden	Switzerland	Turkey	USSR*	UK	Yugoslavia*	Unidentified + other areas	Total
Austria	151	9	–	92	–	66	–	39	54	–	33	–	96	–	5	–	23	–	–	6	–	7	–	–	18	19	72	690
Belgium	–	198	–	–	–	50	–	55	5	–	–	–	–	–	10	–	–	–	–	–	–	–	–	–	41	–	21	380
Bulgaria	–	–	270	21	–	9	–	4	14	10	29	–	14	–	–	–	13	–	98	–	–	–	6	79	5	25	70	667
Czechoslovakia	64	30	4	1266	6	223	–	88	438	–	157	–	67	–	18	–	170	–	98	13	–	–	–	89	73	29	218	2999
Denmark	–	–	–	6	109	23	–	6	28	–	–	–	–	–	–	–	7	–	–	–	–	–	–	–	25	–	26	232
FRG	23	110	–	126	12	1160	–	211	265	–	14	–	41	7	57	–	32	–	6	19	–	14	–	18	165	8	192	2474
Finland	–	–	–	12	8	31	260	10	43	–	7	–	4	–	5	–	27	–	–	–	37	–	–	270	32	2	132	908
France	–	25	–	22	8	203	–	1232	45	–	–	–	70	6	30	6	7	–	6	201	–	18	–	18	226	3	398	2568
GDR	–	–	–	146	8	176	–	–	1117	–	10	–	6	–	16	–	41	–	–	–	–	–	–	–	51	–	62	1720
Greece	–	60	–	90	12	6	–	6	7	137	12	–	24	–	–	–	4	–	23	6	–	–	9	24	7	15	84	519
Hungary	25	–	–	120	–	27	–	–	40	–	448	–	46	–	–	–	37	–	45	–	–	–	–	24	28	42	48	923
Ireland	–	–	–	–	–	–	–	4	–	–	–	96	–	–	–	–	–	–	–	–	–	–	–	–	28	–	62	190
Italy	17	6	–	32	–	46	–	104	24	–	25	–	1357	4	–	–	10	–	8	52	–	16	–	13	20	36	223	1989
Luxembourg	–	–	–	–	–	4	–	6	–	–	–	–	–	4	–	–	–	–	–	–	–	–	–	–	–	–	–	14
Netherlands	–	47	–	5	–	93	–	29	13	–	5	–	–	–	99	–	18	–	–	6	–	–	–	–	62	–	26	374
Norway	–	12	–	21	16	41	7	–	50	–	–	–	31	–	7	61	18	–	–	–	26	–	–	50	92	–	178	606
Poland	20	21	5	356	22	165	–	51	478	–	99	–	–	–	17	–	1012	–	37	6	10	–	–	215	73	22	185	2825
Portugal	–	–	–	–	–	–	–	14	–	–	–	–	–	–	–	–	–	72	–	52	–	–	–	–	–	–	70	194
Rumania	10	–	53	97	–	33	–	14	58	6	164	–	46	–	–	–	68	35	822	–	–	–	9	301	13	64	144	1902
Spain	–	6	–	–	–	41	–	74	9	–	–	–	4	–	7	–	–	–	–	1121	–	–	–	–	34	–	266	1597
Sweden	–	15	–	47	44	72	34	29	95	–	12	–	7	–	12	24	55	–	8	6	218	–	–	152	80	4	271	1179
Switzerland	–	–	–	5	–	27	–	45	7	–	–	–	81	–	–	–	9	–	–	9	–	32	–	–	11	–	41	257
Turkey	–	–	49	13	–	10	–	7	13	29	16	–	22	4	–	–	9	–	35	6	–	–	360	107	–	12	190	880
USSR	41	53	122	496	69	393	168	131	636	32	367	11	161	4	57	15	691	–	587	31	100	7	92	22674	236	115	264	27553*
UK	–	20	–	11	–	43	–	53	24	–	4	36	4	–	12	–	9	–	8	31	–	–	–	–	1545	–	173	1936
Yugoslavia	41	6	57	99	–	46	–	39	56	9	166	–	224	–	5	–	37	11	74	20	–	21	–	50	15	308	209	1461*
Other areas	38	149	150	276	144	522	101	578	481	17	156	106	765	9	133	34	235	33	216	537	119	174	–	1416	1398	120	281	8309
Total	430	810	770	3370	450	3510	570	2890	4000	340	1720	260	3070	30	490	140	2500	140	2000	2090	510	120	650	25500	4250	830	3906	65346

* Cases which are not within the estimates given in reference 20. However the proportions of emissions deposited in each country are the same.

Sources: see text.

Greece (25 per cent). Having said that, their contributions do not affect Western Europe as much as Western Europe's own output, because of the direction of the prevailing winds: twice as much SO_2 pollution arrives in Norway, for example, from the West than from the East.

Which brings the discussion to our own doorstep. Highton and Chadwick's chart shows that the 19 countries in Western Europe and Scandinavia produced 21.6 million tonnes of sulphur dioxide. Three-quarters of this came from five nations — Spain (2,090,000 tonnes), France (2,890,000), Italy (3,070,000), Federal Republic of Germany (3,510,000) and the UK (4,250,000). This last figure is surprising: considering that the UK has a population similar to France's, and less than that of Germany or Italy, the overwhelming British lead in SO_2 pollution takes some explaining. The figure is made even more remarkable by the fact that throughout the 1970s the UK regularly produced 5 million tonnes or more of sulphur dioxide.[7] The more respectable figure that occurs nowadays is an indication of industrial inactivity caused by the economic squeeze in the UK, rather than a deliberate move towards cleaning our industrial emissions.[8]

The diplomacy of international pollution
The five major West European polluters are, with the exception of West Germany, arranged along the southern or western edge of Europe. They get large inputs of clean air from the sea, and although experiencing pollution in deterioration of buildings and damage to health, they have not felt the acidification crisis that has gripped the countries of Scandinavia and central Europe. They know about the primary effects of sulphur pollution, the immediate local results, but have not yet seen the secondary consequences of overloading the environment — the large-scale poisoning of trees and lakes. When other countries in Europe — those which have been affected by air pollution, the 'victims' — start talking about the urgent need to cut sulphur emissions, these major polluters do not feel inclined either to agree or to take action. There are plenty of things to spend money on, they argue, without investing in desulphurisation machinery, the worth of which has not been proven, to solve a problem which has not yet been conclusively established.

That, throughout the 1970s, was the position in the 'Who does what to whom?' debate: the polluters, with their supplies of clean air behind them, refusing to be concerned; and the victims (in this period, only Norway and Sweden) using every means at their disposal to push for an international desulphurisation programme—with little success in practical terms.

The picture changed, however, at the beginning of the eighties, when the Germans gradually realised that their forests were dying. Initially they joined the Scandinavians in their call for Europe-wide sulphur output reductions; but by 1983, amid mounting domestic pressure and concern, they announced a unilateral sulphur reduction programme.[9]

This left the other polluting countries in an exposed position. Five nations still produced three-quarters of Western Europe's air-borne sulphur dioxide, but one of them—the second largest polluter—had publicly declared its intention to reduce SO_2 output by around 50 per cent over the next 10 years. The ground was prepared for a diplomatic assault on the other four. The question was, which one should bear the brunt of the attack?

In polluting terms, Spain was the least important of the 'dirty four' because of its geographical position. Its sulphur dioxide contaminated France, Portugal, and Italy, but these were not countries that counted themselves as seriously affected, and in any case Spain's pollution was small compared to that of the others.

For similar reasons Italy was discounted. It produced a large amount of sulphur dioxide, true, but its mediterranean winds meant that much of this was blown to the south-east; and the Alps formed some sort of barrier between the Italians and the most seriously affected countries. This left two polluters—France and the UK. Britain was the most logical target because of its long tradition of SO_2 emission, the fact that it was the largest SO_2 producer in Western Europe, and it was the area's biggest exporter of the pollution to Scandinavia and Europe. So in 1983, the pressure began.

As late as June 1983, there was very little public awareness in the UK of what the phrase 'acid rain' even meant. Internationally, there had been a few initiatives. The UK had signed the 1979 Geneva Convention, a somewhat toothless document which deprecated international air pollution without committing the signatories to any firm action.[10] The UK had taken part in the 1982 Stockholm Conference on the Acidification of the Environment, which called for a reduction of sulphur emissions.[11] And in March 1983, Mrs Thatcher even put her name to an EEC document expressing concern about the effects of air pollution on European forests.[12] Without domestic concern, however, these international gestures were merely diplomatic politeness. At the beginning of 1983, a UK Department of Environment recommendation to start a modest desulphurisation programme was shelved[13] after lobbying from the Treasury and the Electricity Board — the government could see no political reasons for taking action.

International pressure increased in June 1983 at the United Nations Economic Commission for Europe (the administering body for the Geneva Convention). Nine countries proposed or supported the idea of a 30 per cent reduction in SO_2 emissions over the next 10 years — Sweden, Norway, Denmark, Finland, West Germany, Austria, Switzerland, the Netherlands and Canada. The USSR agreed in principle to a 30 per cent reduction in its SO_2 exports over the next 12 years. The UK, France, and Italy blocked the proposal but, suitably embarrassed, promised 'effective reductions'.[14]

Shortly afterwards the issue became an important one in the UK. Up to this time, it had largely been ignored by the media (with the notable exception of *New Scientist*), but a Royal Society meeting on the subject, a free press trip to Norway for 20 journalists (paid for by the Norwegian Stop Acid Rain campaign) and controversy over a research project to investigate acid rain to be sponsored by the Electricity Board (was it real scientific research, or a blatant attempt to buy time instead of taking action?) made the newspapers amenable to acid rain stories. The discovery that acid deposition was killing lakes in Scotland and Wales[15] transformed the subject into an interesting, if not particularly pressing, domestic issue.

Meanwhile, the diplomatic pressures persisted. The Department of Environment was finding its international negotiations being compromised by hostile reactions to Britain's stance on desulphurisation:[16] the EEC, pushed by West Germany, prepared a draft directive[17] for discussion in June 1984, with the intention of reducing SO_2 emissions from large plants by 60 per cent—on balance a larger reduction than that proposed in the aborted UNECE discussions; and the UK environmental organisations, alerted by contacts overseas, began a concerted campaign to identify Britain as the largest polluter in Western Europe. This campaign received an unexpected boost in February 1984 when France yielded to international representations and announced a 50 per cent reduction of SO_2 emissions between 1980 and 1990.[18] Britain's diplomatic position, previously exposed, became embarrassingly vulnerable, and government statements in the House of Commons began to have more urgency: from the complacency of mid-1982 ('This problem is not nearly as severe as the media make out')[19] to a more earnest, concerned tone by early 1984 ('I vigorously refute any charge of complacency... My department is stepping up substantially the amount of research in this sphere').[20]

The Nordic nations took the unprecedented step of publicly appealing to the UK to move to a 30 per cent reduction of SO_2 on 1 March 1984: this was followed three weeks later by the formation of the 'Club of Thirty' in Ottawa, in which the nine nations who had supported the June 1983 UNECE proposals, plus France as a newly desulphurising nation, pledged themselves to a 30 per cent SO_2 reduction over the next ten years, with the intention of also reducing NO_x. The UK and the USA were identified at these talks as major targets for diplomatic action, and representations were made at governmental levels. In the UK, the domestic campaign against acid rain was not pointed merely at the government, however. It had identified, and was intent on severely embarrassing, the UK's most flagrant SO_2 polluter—the Central Electricity Generating Board (CEGB).

Power station pollution

The sulphur pollution of Europe is caused by power stations and industry. The proportion of pollutant emitted by each sector varies according to the energy needs and industrial activities of individual countries. In West Germany, for example, 56 per cent of SO_2 comes from power stations, 28 per cent from industry, and the rest from small users.[21] In the Netherlands, three-quarters of the nation's sulphur dioxide is produced by power stations and industry, a proportion generally reflected throughout the EEC, with the power stations being by far the largest single emitters.[22]

This is very much the situation in the UK. The trend has been away from the domestic use of coal and oil, and to compensate for this, electricity output has increased to meet escalating demand, with extra production fuelled by large quantities of coal. The result: in 1960, power stations contributed about one-third of Britain's SO_2 pollution; in 1970, about 45 per cent; and in 1983, 65 per cent.[23] Nowadays, two-thirds of our enormous production of SO_2 escapes through the tall stacks of the electricity utilities rather than at lower levels. This leaves the local environment freer of pollution, but increases the risk of long-range travel for the pollutants (and thus of their falling as acid rain).

An interesting consequence of this change of energy use in the UK is that *one* organisation—the Central Electricity Generating Board—is now responsible for three-fifths of Britain's sulphur dioxide output. The CEGB supplies about 90 per cent of airborne SO_2 from UK power stations, a total of 2.4 million tonnes a year[24]—more than that produced by any other Western European organisation. In fact, its airborne SO_2 output is larger than that of several European *countries*, including Austria, Belgium, Bulgaria, Denmark, Finland, Greece, Hungary, Ireland, Holland, Norway, Portugal, Rumania, Spain, Sweden, and Switzerland. If the CEGB were to declare UDI and take on national status, it would be the fifth largest SO_2 polluter of Western Europe. The implications of this are not lost on the UK environmental movement, who angled their 1984 campaign to point out the size and persistence of the CEGB's poisoning of the air. The CEGB's environmental credibility, damaged by a series of radiation leaks and blunders at Windscale (not owned by them, but associated in the public mind with nuclear reprocessing from power stations), suffered further deterioration because of the organisation's policies on acid rain.

· As far as NO_x pollution is concerned, power stations also bear a large portion of the blame, though the picture varies from country to country. About 50 per cent of Europe's NO_x emissions come from traffic, with most of the remainder coming from power stations and space heating.[25] In the UK though, 45 per cent of NO_x comes from power stations, with 26 per cent from traffic and most of the rest from industry.[26] National NO_x surveys for individual countries are still at a rudimentary stage,[27] but in the UK at least, the largest producer of SO_2 is also the largest emitter of NO_x.

Technology and costs

The technology exists to clean 95 per cent of the sulphur dioxide from smoke[28] and 90 per cent of NO_x.[29] SO_2 control has already been instituted in large installations in certain countries, notably Japan, Germany, and the USA. The systems for cleaning smoke are described in Appendix B, but can be summarised here as: (a) taking the sulphur out of the fuel before it is burnt; (b) changing the furnace in which combustion takes place; and (c) filtering the smoke as it leaves the power station or factory chimney. In practical terms, system (b) is ruled out because it is an option for future plant and not really suitable for existing facilities. System (b), although useful for partially reducing SO_2 emissions, has not yet achieved a large enough sulphur extraction rate to be the basis of a national desulphurisation policy. This leaves system (c).

Cleaning the sulphur out of smoke as it rises up the chimney is called Flue Gas Desulphurisation (FGD) and has been in useful operation for some time. In fact, Battersea Power Station, built in 1929, was the first installation in the world to have a succesfully operating FGD system, cleaning 80−90 per cent of sulphur from the plant's smoke for most of its working life. In the UK, all large modern power stations are required at the planning stage to provide enough space on site for FGD to be installed if necessary.[30] In practical and technological terms, the fitting of FGD is perfectly feasible: Britain's sulphur dioxide output could be halved in the space of a few years. Whether this will happen or not is a question of politics and money.

It would not be too expensive to reduce the UK's output of SO$_2$ by 30 per cent, a target developed in the 1983 UNECE negotiations and requested in 1984 diplomatic representations to the UK government.[31] The reduction could be achieved most simply by halving the CEGB's production of sulphur dioxide, a process which on the CEGB's own figures would add 6 per cent to electricity prices over a period of ten years[32]—an annual increase of 0.6 per cent. To put these additional costs into perspective, domestic electricity prices increased 4.6 per cent per year between 1974 and 1982—and this was a real increase, after inflation had been taken into account. Industrial electricity prices increased over the same period by 6 per cent per year in real terms.[33] In comparison with these figures, desulphurisation costs are not by any means excessive—even less so when independent analyses are considered, which put the likely electricity price increases over 10 years at 4—5 per cent,[34] or, in another study, at 3—4 per cent.[35]

In cash terms, the cost of this programme would be—again using the CEGB's own figures—an initial £1,500 million with annual operational costs of £250 million.[36] The estimated costs of building a nuclear power station at Sizewell are £1,184 million.[37] If this shows the overruns experienced at four of the CEGB's other nuclear power stations,[38] this would increase to a minimum £1,500 million—the same price as a desulphurisation scheme. The operational price of desulphurisation can be compared to the estimated annual costs of damage to crops in the UK from sulphur pollution—£200 million—and this ignores those costs arising from corrosion of buildings, deterioration of lakes, health damage, and possible destruction of forests. The OECD has calculated that a desulphurisation scheme such as this would have roughly comparable costs and benefits in terms of environmental improvement:[39] this would appear to be the case as far as the UK is concerned.

Alternative systems of energy
The perceptions driving the acid rain campaign—that a particular fuel use is polluting the air, the earth, and large expanses of water—are the same that motivate the campaign against nuclear

power, with the significant difference that the technology exists to clean fossil-fuel pollution. Techniques to neutralise nuclear waste, or to contain airborne pollution in the aftermath of a nuclear core meltdown have not been developed. For this reason, the environmental movement tends to react strongly against the suggestion that arguments against acid pollution are statements in favour of nuclear power.

Environmentalists have noted that oil is currently too expensive to be considered as a base fuel for electricity production,[40] and that the choice for the next generation of power plants rests at the moment between coal and nuclear fuel. Notwithstanding legitimate environmental criticisms of nuclear power—the problems of by-product storage, potential danger to human health and human life, hidden links between civil nuclear programmes and nuclear weapons, the appalling safety record of the nuclear industry in the UK—there are also pressing economic reasons why coal will be the preferred choice of fuel for the UK in the early twenty-first century. Britain has enough coal reserves to last for the next 300 years,[41] and nuclear power is becoming identified as potentially more expensive than any other available fuel, depending on safety standards applicable and other cost factors.[42] Even Sir Walter Marshall, the strongly pro-nuclear head of the CEGB, is backing down from his 1982 statement that the UK should be building one new reactor a year towards the end of the twentieth century, and now says that the CEGB 'are not locked into a nuclear power construction programme'.[43] Pending the development of acceptable alternative sources of energy—which include wind and wave sources, tidal barrage schemes, combined heat and power systems, and a massive drive for energy conservation—the environmental movement is quite happy to embrace coal as the fuel of the future—but *clean* coal, rather than the present variety.

Technology versus jobs

The comment is sometimes made that, with the level of unemployment in the UK at the moment, there is no justification for installing expensive desulphurisation machinery which will divert investment from job-creating schemes. Or, as the Financial

Controller of a company puts it:

> I've got £100,000 to spend, and I have to justify the
> expense to my Managing Director. I have the choice
> between installing machinery that will provide jobs and a
> good return of capital, or sticking some new-fangled invention
> up the factory chimney with the long-term intention of saving
> a few fishes' lives in Sweden. What do you expect me to do?

In terms of his company's balance sheet, he is correct, though
an equally valid argument could be made that other people are
currently paying for the firm's chosen methods of waste
disposal, and there are no reasons why the company should not
bear these costs itself. In national terms, however, there are
several reasons why a clean air policy should be actively
pursued:

- the costs of desulphurisation would be offset by savings in
  environmental deterioration, corrosion of materials, and
  damage to health;
- the development of a realistic energy conservation
  policy—a central plank in the environmentalists' plans for
  a reduction of SO_2 emissions—would save, in the view of
  a Conservative Energy Minister, £7,000,000-worth of
  wasted energy a year;[44]
- desulphurisation expertise will be a major money-earner if
  the EEC moves to a Community-wide sulphur reduction
  programme[45] (certain to happen sooner or later), and those
  countries which have developed their own systems will
  have a disproportionate advantage in the billion-dollar
  market. Places such as the UK, with no practical
  experience in the field, would be forced to import
  machinery or operate under licence—thus losing investment capital and hundreds if not thousands of job
  opportunities.

Jobs, in fact, present another argument against nuclear power.
Every nuclear station opened costs 3,500 mining jobs, with a
similar number of jobs lost indirectly. A programme of 10

Pressurised Water Reactors, such as that suggested for Sizewell, would displace 70,000 people.[46] A programme of desulphurisation, carried out at a fraction of the cost, could generate jobs and provide useful opportunites for foreign currency earnings.

NO_x

The technology for reducing NO_x is less expensive than systems for cutting SO_2, costing up to 6 per cent of electricity generating cost compared with $5-15$ per cent for SO_2.[47] (These are not the costs passed on to the consumer, as the price of elec- tricity is assessed from several factors, only one of which is the actual cost of generation.)

The acid rain campaign in the UK is not pushing for NO_x filters to be installed on large plant, though this is very likely to be an aim in the near future. In central Europe, however, where ozone (created from NO_x precursors) is regarded as a signifi- cant problem, moves towards lead-free petrol with the intention of fitting NO_x-reducing tri-catalytic converters to vehicle exhausts, have already been implemented. Installation of NO_x filters to stationary sources is a probable next step: West Germany, for example, has just launched a programme to reduce NO_x from large stationary sources by 40 per cent, and this has been incor- porated in draft EEC legislation.

Two practical examples

In any debate over an environmental issue, the practicality of the solution is as much an assessment of political attitude as of objec- tive analysis. In 1866 for example, there was a UK cholera epidemic when the Thames was an open sewer and 14,000 people died. At that time sewage treatment had hardly begun and *The Times* declared: 'We prefer to take our chances of cholera than be bullied into health.'[48] This philosophy was also expressed by Harold Macmillan, Minister of Housing and Local Government, shortly after the 1952 London smog when 4,000 died:

Today everybody expects the government to solve every problem. It is a symptom of the Welfare State ... For some reason or other 'smog' has captured the imagination of the press and the people ... Ridiculous as it appears at first sight I would suggest that we form a committee ... We cannot do very much, but we can seem to be very busy—and that is half the battle nowadays.[49]

The same political statements are surfacing at the moment about the impracticality of desulphurisation and denitrification programmes; in essence, that the ideas are worthwhile but not at all realistic in today's economic circumstances. There *are* places, however, where air pollution control schemes have already been implemented.

Japan Over the past twenty years, Japan has experienced a startling rate of economic growth, twice as rapid as the OECD average.[50] This has been accompanied by the accumulation of large industrial complexes in the cities, and by massive migration from rural to urban areas. The population of Japan is densely grouped round industrial emission sources. Initially, there were very severe pollution episodes in some places, and it soon became apparent that stringent emission regulations were necessary to protect the health of the populace.

The Air Pollution Control Law was passed in 1968 and amended in 1970 and 1974.[51] Under this law, restrictions are imposed on the emission of sulphur oxides, nitrogen oxides, particulates, dust, production of other noxious substances from industrial plant and emissions from motor vehicles. This is achieved for sulphur through the use of the 'K Value Control', a measure for limiting the amount of sulphur oxides reaching the ground, rather than just controlling emission. The reason for this, say the Japanese, is that the K system more readily allows a direct comparison of effects on the environment. Instead of working from the chimney and seeing what the results are, they work from optimum ground levels and try to find ways of achieving them.

Originally, maximum emissions were established by central government on a stack-by-stack basis, although these standards could be tightened by regional directives if it was thought to be necessary. Then, under 'total mass emission control', local authorities indicated the overall amount of sulphur emission that they would tolerate in each regulated area, drew up programmes to reduce the volume of airborne effluent to the levels indicated, and allocated emission 'rations' on a plant-by-plant basis. In 1981, the same system was introduced for oxides of nitrogen.[52]

Central government assisted by a system of positive measures. It reduced import duties on low-sulphur fuel oil, and when this became more difficult to get after the 1973 oil crisis, government funds were advanced for the installation of hydro-desulphurisers and flue-gas desulphurisation. The former reduces the sulphur content of crude oil: using this system, a crude oil sulphur content of 3 per cent could be reduced to 1.5 or 1.7 per cent. The latter cleans the sulphur out of smoke as it is going up the chimney. Between 1970 and 1975, over a thousand of these systems were installed, with a desulphurising efficiency of 85−90 per cent.[53] The combination of these measures, and the government's determination that emissions *were* going to be cleaned up, had a definite result. In the 10 years from 1968 to 1978, environmental sulphur concentrations were reduced to one-third of their previous levels, even though the consumption of petroleum in the same period increased threefold.[54] The sulphur reduction is impressive by any standards, but it is even more so given the enormous increase in fuel consumed in Japan at this time.

The secret of this Japanese success is that they sincerely *wanted* to reduce their sulphur and nitrogen pollution−so they looked at the technical possibilities and worked out the fastest way of achieving their environmental goals. This process also involved adapting to outside events. Their reliance on fuel oil was shaken by the oil crises of 1973 and 1979, and they are now planning a move towards coal. But as they stated to the OECD in 1982: 'It is extremely important to incorporate environmental considerations into the planning of coal-fired plants, coal centres and other coal-related projects.'[55]

Their pollution control has not been achieved without a certain amount of expense and pain. There is a different philosophy at work here, however: as mentioned on page 44, if people pollute the air in Japan they have to pay compensation to those who suffer from the effects of this atmospheric poisoning. The Japanese recognise the destructive effects of sulphur and nitrogen emissions, and treat them accordingly. A report from a member of the Environmental Protection Policy Division in Japan discussed industrial pressures on the government over the national abatement measures for nitrogen dioxide:

The Japanese standard for NO_2 is stricter than those in other countries and is under attack from several quarters. As the critics argue, standards for NO_2 are without scientific basis and cost vast sums to achieve. The epidemiological research used as the basis of the NO_2 standards is not too convincing and it seems that there is no proof that NO_2 only slightly above a specified level is harmful to the health. But the contrary (that is, that NO_2 at such concentrations is harmless to the health) has not been strictly proven, either.[56]

It would be extremely unusual to see such a report from the UK Department of the Environment. The attitude here acknowledges that there is an area of doubt in the controversy of clean air versus NO_2 pollution at this concentration; but in the absence of conclusive evidence either way, the scales are tipped in favour of clean air. Governmental policy in the UK still works from the other end: it waits until the pollutant produces disastrous effects before it starts to take action. As another Japanese scientist noted, in 1978:

Japanese environmental standards are internationally by far the most stringent (several times more so than the USA, West Germany, or other countries). (Many countries, including Britain and France, are without environmental standards altogether. Norway and the EEC bloc are proposing standards that are at the American level.) Environmental concentrations of pollutants in the principal Japanese cities are about the same as those in Europe and America, but these countries simply ignore pollution at this level.[57]

USA The Bruce Mansfield electricity generating plant in Pennsylvania is one of the largest coal-fired plants in the USA, and became operational in 1981 at a cost of $1.3 billion (£1 billion, 1984 prices). It has a generating capacity of 2.36 million kilowatts, enough to serve the needs of 500,000 homes, and can consume 24,000 tons of relatively high-sulphur Ohio coal per day. When the product of this combustion process leaves the Mansfield stack, it is 92.1 per cent free of sulphur dioxide—at an additional cost to the customer of an extra 7 per cent on electricity charges.[58] As mentioned before, the current request to the UK is for a 30 per cent reduction in SO_2 emissions.

Systems for delay

The argument over acidic air pollution is complicated by being essentially about time. It is generally agreed that measures to reduce sulphur and nitrogen oxides will be imposed in Europe and North America sooner or later, but it is in the interests of the electricity lobby to ensure that 'later' is the operative word. The reason for this is that there are a number of smaller power stations still in operation which are relatively more difficult and expensive to clean, and also larger generating stations with 15−30 years of working life left. In the UK, for example, the 10 coal-burning stations that produced most electricity in 1982/83 were all built between 1968 and 1979,[59] and would reasonably be expected to continue functioning until the end of the century. The electricity authorities would prefer to let this generation of plant retire gracefully rather than submit them to desulphurisation programmes, a process which is 20−30 per cent more expensive to fit to existing plant than to build as new.[60] If action were to wait until the 1990s, desulphurisation would be cheaper, easier, and might also be able to take advantage of new technology.

The environmentalists, on the other hand, want action immediately. Acidification damage has multiplied since the 1960s; the number of affected lakes is expected to double by the end of the 1980s;[61] and the effects on forests of continued pollution can at the moment only be guessed at. Ten or fifteen years' delay on smoke-cleaning represents environmental destruction on a potentially enormous scale. What was discovered in 1964 became a Scandinavian problem in 1973 and a European dilemma by 1983.

Another ten years could turn it into a catastrophe.

Governmental attitudes are naturally crucial to this debate. If an administration wants to desulphurise, it will do so regardless of protests from industry. On the other hand, if it feels no particular pressure in this direction, there are plenty of ways in which a desulphurisation programme can be postponed. In any argument based on time, a government has an inherent advantage because there are so many ways in which a proposal can be shelved. A hostile administration can initially ignore an environmental problem, then indulge in diplomatic activity which although useful in public relations terms involves no commitment to action. If pressure persists, lengthy research projects can be launched, which for a minimal investment defer the date of remedial initiatives by five or ten years. Finally, when there are no alternatives left, the government can announce a programme but delay its implementation for a number of plausible reasons.

The UK government is in the middle of the 'research' stage. One enquiry, which was precluded in its terms of references from considering the causes or effects of acidification in the UK, has just reported that domestic acid rain definitely exists;[62] and another, which has limited research objectives and is scheduled to last five years, has just been set in motion.[63] Britain is also putting down markers for the next stage, however. One response to the draft EEC pollution-reduction legislation is to seek special treatment for countries whose energy policies depend in large part on fossil fuels (e.g. the UK): and the CEGB is generally letting it be known that, should a desulphurisation policy be forced upon the UK, it would not be possible in a shorter period than ten years[64] – and this regardless of the fact that the Germans have already pledged themselves to the sulphur reduction programme under discussion, and hope to achieve the bulk of their objectives within five years.[65]

It is one of the unfortunate aspects of the debate over air pollution in the UK that any statement from the CEGB needs to be examined meticulously. It is a major SO_2 emitter which has made no practical attempts to clean up its sulphur dioxide pollution, produces substantially the same amount of SO_2 now as it did ten years ago,[66] and has no intention as yet to curtail its emissions. It is also a politically powerful institution with massive financial resources at its disposal, defending its position by a barrage of public relations statistics and postures, some of which are misleading, some concentrating on one aspect of the truth to the exclusion of the broader picture, and others frankly nonsensical. While not denying the extent of its SO_2 emissions, it rests its defence on an argument that the case against sulphur dioxide production has not been conclusively established and that other factors are responsible for the deterioration of Europe's lakes and forests. A number of statements have been made in the course of this defence, some of which deserve to be examined more closely:

There is no scientific consensus as to the cause of the acidification of lakes.

This is a statement for domestic UK consumption, and is no longer put forward at international scientific conferences because of the reaction it provokes. A number of international organisations, including the United Nations Economic Commission for Europe and the United Nations Environment Programme, have issued appeals for the reduction of sulphur oxide emissions on the grounds that they cause acid rain and the destruction of the environment. On 20 September 1982, for example, a UNECE report specifically stated:

> considering the evidence of ... long-term effects of acidic deposition on the quality of groundwater, with implications for potability (drinking water) and the acidity of streams, rivers, and lakes in sensitive areas ... it is recommended that the reduction of atmospheric pollutant loading, particularly sulphur compounds, receive priority attention.[67]

The 1982 Stockholm Conference on the Acidification of the Environment, at which the CEGB was represented, concluded that, 'For an increase in acidic sulphur deposition there will either be an increase in soil acidity or water acidity, or a mixture of the two.'[68] In the conclusions to the 1983 EEC Conference on Acid Deposition at Karlsruhe, Germany, it is stated:

The signals are loud and clear: deterioration of buildings, corrosion of metal structures; acidification of lakes; dying off of forests ... Measures have to be taken now [which include] a harmonised European policy ... to establish national emission levels, and to reduce those levels in time, e.g. 30 per cent in five years.[69]

The director of the United Nations Environment Programme reported in 1983, 'The rain ... mixes in the air with pollution from burning fossil fuels ... and brings down dilute sulphuric and nitric acid. This is killing fish and other water life.'[70]

There *is* a scientific controversy, but this is concerned with the exact nature of the pollutant that is killing Europe's forests. Is it ozone or acidic stress? Even within this discussion, however, there is general agreement that the problem is caused by air pollution, and remedial measures will probably have to be taken to reduce *both* ozone production and acidic emissions.

There is no proportional relationship between the emission of sulphur oxides and acidic deposition.
This is otherwise known as the 'linearity' issue, and boils down to a statement that sulphur emissions turn into sulphuric acid only in certain meteorological conditions; these conditions do not always occur; so the transformation will not always take place. The implications of this are that a given reduction of emissions will not result in a proportional reduction of acid rain — so all the expense involved in desulphurisation might not be justified.

The theory is weak because it does not attempt to explain what happens to untransformed SO_2 — does it disappear, or what? — but it has been given pride of place by the CEGB, and even appears in their 1982/83 *Annual Report*:

In clear sunny conditions the oxidation of sulphur dioxide occurs very slowly and is strongly influenced by the presence of other pollutants such as hydrocarbons in the air. The potential conversion of chimney emissions to acidic material in the atmosphere is believed to depend critically upon both the availability of photochemically produced oxidants and the meteorological conditions prevailing at the time. In such circumstances, the acidity of rain in Europe cannot be expected to respond proportionately to changes in the amount of sulphur dioxide emitted. This has been drawn to the attention of the government because of its important policy implications.[71]

In 1983, the theory was discredited in a number of reports. The National Research Council (NRC), USA, looked at emission and deposition patterns over a large area and concluded that:

there is no evidence that the relationship between emissions and depositions in north-eastern North America is substantially non-linear when averaged over a period of a year and over dimensions of the order of a million square kilometres ... if the emissions of sulphur dioxide from all sources in this region were reduced by the same fraction, the result would be a corresponding fractional reduction in deposition.[72]

This research was reported to the September 1983 UK Royal Society hearing on acid rain,[73] and was supported in the summing-up of the EEC hearing later in the month:

Is there a linear relationship between emissions and depositions? ... Generally speaking, on a global scale in the long-term average there must be such a linear relationship. A rough linear relationship ... is also given for a large European area in the long-term average (for example one year).[74]

On 1 December 1983, the relevant working group of the UK Watt Committee on Energy reported to a meeting on acid rain that this analysis had also been confirmed by their research.[75] CEGB scientists were present at the Royal Society meeting, the EEC hearing, and the Watt Committee proceedings; yet on 15 December 1983, a CEGB representative stated to a conference in University College, London, that the relationship between emissions and depositions is non-linear. It will be interesting to see if this theory appears again in the CEGB's 1983/84 report.

The UK has already played its part by reducing its sulphur oxide emissions by 25 per cent in the last ten years. No other country in Europe has reduced its output to pre-second world war levels as we have.

The facts here are true, but the gloss is misleading. The statement skates over the point that the UK was the largest SO_2 polluter in Western Europe before the war, continued to be so afterwards, and still holds that dubious position now, even after its SO_2 reduction. Also, the major drop in SO_2 output occurred in 1980 and 1981, when British energy demand fell by 13 per cent and oil prices rose.[76] There is no planning in the UK for what would happen if there was an economic revival. Our sulphur dioxide output would probably increase to the level of the early 1970s.[77] Official self-congratulation over our emission reduction does not mention the fact that we have now moved from a position of being the worst SO_2 producer in Western Europe, to being the worst SO_2 producer in Western Europe.

A similar point is sometimes made here that the UK reduction has achieved no results in Scandinavian environmental improvement, so why should we aim for more reductions? The answer, of course, is that we have not reduced enough, and that other European countries have stepped up their SO_2 output, to a total that is at best 100 per cent larger than forty years ago.[78] The solution will have to be a Europe-wide reduction and the UK, as the major polluter in north-west Europe, has a responsibility to accept and an example to set.

The CEGB is extremely concerned about acid rain, conducts £2 million-worth of research every year into the problem, and has just sponsored a major research project which will be examining the phenomenon in Scandinavia.

Again, the facts are true, but the interpretation needs to be looked at with care. The CEGB's position, as explained to me by two of its representatives, is that of devil's advocate. It sees European scientists linking sulphur emissions and acid deposition, and its duty to the electricity consumers of Britain means that these theories and arguments should be tested and assessed. The CEGB's scientific research is framed so that it does not ask the question, 'What causes acid rain?' (the answer might come back, 'We do.') but rather, 'What, apart from sulphur oxide emissions, could cause acid damage to the environment?'

In the same way that tobacco industry research into lung cancer will produce all kinds of valid contributors to this ailment—hereditary factors, work environment, air pollution, for example—the CEGB work has pointed out that there are other causes of acidification. If you plant coniferous trees near a vulnerable lake, or change the drainage system through a nearby peat bog, for example, this could turn the lake acid. Similarly, overuse of nitrogen-based fertilisers, or the felling of large numbers of trees, can change the acid balance of groundwaters. And the CEGB has pointed out that the acid input from industrially-caused emissions is small compared to the vast natural processes at work in forest and lake systems.

On the other hand, however, none of the CEGB's researches has been able to demonstrate *why* enormous parts of Europe and Scandinavia are now suffering acid damage. The acidifying natural processes have been occurring for hundreds of years; why is all this destruction happening now? The Scandinavians have pointed out that the quantities of industrially-based acidity are large enough to tip the delicate balance that has hitherto existed in nature; but the CEGB refuses to confront this question, and keeps prodding at the edges of the acidification debate.

In domestic terms, the large amounts of time and money that the power lobby has devoted to distractors has distorted the discussion on acidification in the UK. British research in this area has been criticised for being obsolete and 'woefully inadequate'.[79] The participation of a major research effort which is trying to disprove rather than test a point has exacerbated the lopsidedness of the UK approach to acid rain.

As for the 'new' research project: this was launched with a great fanfare in September 1983, when the CEGB and the NCB announced that they would each pay the Royal Society £500,000 per year to carry out research into acidification in Scandinavia, over a period of five years. Norwegian scientists have variously described the project as 'dirty money' and 'rediscovering the wheel',[80] saying that the areas to be researched have been extensively examined already, and that the project is nothing more than a crude attempt to buy time. Sir Walter Marshall, Chairman of the CEGB, did nothing to allay these criticisms when he held a press conference to launch the research and announced that the CEGB would not desulphurise any power stations until the project had completed its five-year run. The environmentalist lobby has since pointed out that the CEGB has an income of £19 million a day, or roughly £800,000 an hour. For a cost equivalent to three hours' income, the CEGB is appearing to postpone a billion-pound desulphurisation scheme for five years. An interesting use of science, they argue, to serve a political end. (The Royal Society project had its first meeting in November 1983, and another one in February 1984: 'No, I wouldn't say it was a leisurely timetable,' said the press office when questioned. 'These things take time.')

More research is needed before action is taken.
The Canadians pointed out in October 1982 that, up to then, 3,000 studies had been carried out into every aspect of acidification.[81] Canadian Minister of the Environment, John Roberts, subsequently made a speech in which he savaged those countries who were still calling for more research:

> It's a bit like saying it looks like a skunk, it walks like a skunk, and it's stinking the house up like a skunk, but we are not prepared to commit ourselves that it is a skunk without four more years of research ... We do have enough information to act, it's not a matter of science any more, it's a matter of political will. We have reached the point where a decision to stall and drag our feet on the pretext that we need more research is, in fact, a decision to do nothing.[82]

When examined, these calls for research have a certain element of hypocrisy. In February 1983, for example, the US ambassador to Canada was explaining the need to find out more about acidification before action is taken: 'We need to find out more about this dragon before we go out to slay it.'[83] If he had read the *New York Times* seven months earlier, however, he would have found out that one of the best US research institutes—the National Academy of Sciences—had had its acid-rain research funding from the government cut off, because a report it issued in 1981 had been too alarmist about the acidification of North America.[84]

Similarly in the UK: while British officials were pleading at the June 1982 Stockholm Conference on Acidification of the Environment for more study of acidification, they neglected to mention that the UK government had moved two months earlier to *cut* the amount being spent on research into this area. Britain's foremost centre for this work— the Institute for Terrestrial Ecology—had been presented with a £120,000 reduction in grant.[85]

The calls for research need to be taken with a pinch of salt and examined for the political motives underlying them. There is always the need for more study of an environmental crisis of this scale; but it should be scientific enquiry for the purpose of finding information, rather than a convenient device for postponing action.

The 30 per cent reduction figure has been plucked out of the air, and has not been produced by scientific research: there is no point in desulphurising until we know that it will produce results. The first part of the sentence is correct: some environmentalists have attacked the call for a 30 per cent reduction in SO_2 emissions, arguing that a much larger cut—maybe as much as 70 per cent—is necessary to properly protect the forests of Europe. The reason for this is that, until the reductions are made, their effects cannot be gauged. For a pollution as vast as this, theoretical forecasts of clean-up programmes may need to be changed as the results of desulphurisation and denitrification are observed.

On the other hand, the CEGB position is logically inconsistent. In effect, they are saying that they will continue to produce their pollutants until science finds out what reductions are necessary to protect the environment; but as science cannot find out this protection level without observing the effects of a pollutant reduction, there is no point in their taking action. Other observers might phrase the argument differently.

Liming is a more efficient method for countering acidification than reduction of emissions.

The argument here ventures into the nonsensical, but this statement has been made in all seriousness by the CEGB, so it needs to be examined. Sweden has conducted some experiments with dropping lime into its affected lakes, with the intention of reducing the acidity of the water. The result is that some lakes, in which the fish would have died, have been able to continue to support fish life. The Swedes spent $10 million on liming in 1983,[86] but they are very unwilling to regard this process as anything but a stop-gap measure. 'Liming is a complement [to reduction of emissions], as medical treatment to a sick patient, but certainly not an alternative.'[87]

There are a number of reasons why liming is an inadequate solution to acidification. Firstly, it does not cure the problem, but temporarily alleviates it. Poisonous metals still move from the nearby catchment area into the acidified water, where they stay on the bottom of the lake while the effects of the lime are operative. As soon as these effects begin to wear off, however, the metals begin to circulate again in the newly acidified water, and are joined by other mobilised metals from the surrounding area. To stop this occurring, there is a need for the lime effects to be continually in operation; this implies a liming programme every four or five years.

Secondly, liming does not always work. It does not always prevent, for example, an acid rush from killing a population of fish in the early spring, if the rush is strong enough. Liming is not a solution for quick-running streams, as the lime disappears too quickly for long-lasting effects; and it is difficult to apply to waters which are inaccessible, or to the thousands of small streams and brooks that exist. A liming programme to counter acidification of the land would need a hundred-fold increase in the amount of lime required in order to give acceptable leaching rates;[88] and complications have been experienced in the use of lime to neutralise drinking water — its effect drops off dramatically where the water runs through a long distribution network or stands for long periods in lead tanks.[89]

Thirdly, the proposition itself is preposterous. To take an

analogous situation, if the hotel-keepers of Brighton were faced with the long-term pollution of their beaches by a continual oil spillage from France, there would be uproar. It would be no consolation that detergents adequately clean the beaches, or that pontoons anchored half a mile off-shore would protect swimmers in certain areas from contact with the oil. The people of Brighton would point out, rightly, that the pollution was coming from France and it was up to the French to do something about it; and any talk about cleaning up the pollution with detergents would be insulting and inflammatory nonsense. That is the strength of the Swedish reaction to the proposals about liming, and it is a measure of the poverty of the CEGB's position that this option is being seriously proposed.

So, there is more politics than science in much of the debate about acidification and, as a rule of thumb, a country's position in the argument can be determined by whether or not it feels itself to be affected by air pollution. If it is a victim of acid destruction, it will want to do something to reduce industrial emissions of sulphur and nitrogen oxides. If it is not a victim, it will want a (preferably lengthy) programme of research into the phenomenon before committing itself to action. There is one ray of hope for environmentalists: in 1979, the West Germans were as strong on procrastination as the British, and were accused at more than one conference of dragging their feet. As soon as they perceived that the Black Forest was being devastated by air pollution, however, they changed their minds and their stance, and have now become a very powerful force within the EEC for desulphurisation programmes. The West German 'turnaround' occurred in less than three years, and carried even the most right-wing politicians with it. There is a strong possibility that other countries may experience a change of heart equally as rapid. The aim of the environmentalists is to encourage this change of heart before it is too late for the lakes and forests of Europe.

4. Acid countries

The purpose of this chapter is to provide a country-by-country round-up of acidic air pollution. No distinction is attempted between damage caused by acid rain, dry deposition, acid mists, ozone damage, etc. on the grounds that the root causes of the pollution are the same, even if the transport mechanisms may vary. Damage to materials is not listed, and can be found on pages 35 to 37. The information, written in March 1984, is not intended to be a comprehensive analysis, either of countries or of damage within each country, but rather an initial assessment of research so far. The USA and Canada have been included because of the effects of acid damage in North America, and also because of parallels between the European and American experiences of air pollution.

Sweden

Research
Details of Swedish research are given on pages 115 to 117.

Effects
There are about 90,000 medium to large-sized lakes in Sweden, with a total area of 3.8 million hectares (15,000 square miles), occupying 9 per cent of the country. The acidification of freshwaters started about fifty years ago, but the clearest signs began appearing in the 1950s and 1960s. Today, 18,000 lakes (corresponding to 10 per cent of the total lake area) have a pH below 5.5.[1] Increases in lake acidity of 10−100 times has been documented in the most acidified areas. About 4,000 of the lakes

are what the acid rain campaign would call 'dead', and in about 9,000 — chiefly in central and southern Sweden — fish stocks show more or less serious acidification damage, ranging from minor upsets in life cycle to outright eradication of species.[2]

Brooks and small rivers of a total length of 90,000 kilometres (56,000 miles) have a pH below 6, and 19,100 kilometres (12,000 miles) below 5. In a quarter of the country — in an area four-fifths the size of England — the majority of waters reach unacceptable acidity during some time of the year.[3]

Fishing in more than 100 lakes is banned because of the danger of mercury poisoning.[4] The Swedish authorities have also warned people not to eat certain kinds of fish, or the kidneys and livers of certain game, because of their poisonous heavy-metal content.[5] Very high levels of aluminium have been detected in pied flycatchers nesting on the shores of acidic streams and lakes.[6] Female birds have laid eggs with thin or no shells and have eventually died as a result.[7] Acid well-water, drunk by nearly half a million Swedes, has up to a hundred times more cadmium in it than is normal; the acid well-water also dissolves copper pipes used in most homes and is suspected of causing diarrhoea in children.[8]

Keith Norton, Ontario's Environment Minister, visited Scandinavia and reported that concentrations of cadmium, mercury, copper, and lead had reached toxic levels at a Swedish hospital which relied on an acidified body of water for its needs. 'In at least three documented cases', he said, 'three kidney patients died from a high aluminium content in their bodies, which they received from the water used in the hospital's dialysis machines.'[9] Hospitals in Sweden are now required to check all dialysis water for aluminium content before it is used.

Though previously it had been believed that forests in Scandinavia were unaffected by air pollution, the Swedish Forestry Commission completed a crash survey in late 1983, with new information supplied by the West Germans as to what symptoms to look out for. Preliminary findings indicate that tree damage has been observed in all nine southern counties surveyed, affecting Norwegian spruce and Scots pine — the two totally dominating species in Sweden, and therefore the ecologically and economically important ones. Mature trees, and those exposed to winds from the south and/or south-west, show symptoms of sickness, and damage is worse where the trees are standing on poor soil. The most affected area is in the west coast: in limited areas in two counties, every second spruce over 60 years old bears the signs of failing health.[10]

The OECD calculated in 1981 that at least $10 million of material damage from acidification is suffered in Sweden each year.[11]

Remedial measures

Swedish emissions of SO_2, which were 900,000 tonnes annually in the early 1970s, and 644,000 tonnes in 1977, will be reduced to less than 350,000 tonnes in 1985. This is being achieved by reducing the maximum sulphur content of fuel oil to 1.0 per cent, the use of low-sulphur coal in small installations, and flue-gas desulphurisation in large ones. In 1987, legislation in relation to sulphur emitted by small installations will be enacted.[12]

The Swedes lime some of the lakes that are becoming acidified, as this helps to keep fish populations alive. In 1982 they spent $6 million on liming, in 1983 $10 million, and the projected budget for 1984 is $13 million.[13] They emphasise, however, that this is a short-term measure, as it does not restrict the movement of poisonous heavy metals, and does not always prevent acid surges from killing fish. 'Liming is like giving artificial respiration to a dying patient. It keeps the person alive for a while, but does nothing to solve the long-term problem.'

Sweden supported the 1983 call at UNECE for an across-the-board reduction of SO_2 emissions by 30 per cent.

Norway

Research

The SNSF programme, running from 1972 to 1980, was one of the largest research projects into acidification in the 1970s (see page 117). State and private research in Norway is still continuing.

Effects

There are 230,000 lakes in Norway, a large number of these being located in a high precipitation area, on a geologically vulnerable base. Lakes in more than 1.3 million hectares (5,000 square miles) of southern Norway are practically devoid of fish, and in another 1.9 million hectares (7,500 square miles) of lakeland, the fish stocks are reduced. In the four southernmost counties, more than half the total fish population has been lost in the last forty years.[14] All in all, fish populations in an area the size of Belgium have been destroyed or seriously affected.

The first signs of acidification came at the beginning of the century when large-scale salmon kills occurred in 1911 and 1914. Records of yearly yield from major salmon rivers are available since 1879: the catch for the seven major southern rivers shows a declining trend from 1910 to 1970, when the figure finally dropped to zero.[15]

Deaths of brown trout began to be recorded in the 1920s, and more than 2,000 populations are now lost. The rate of disappearance has been particularly rapid from 1960 onwards and, if this trend persists, more than 80 per cent of the trout populations in the southernmost counties will vanish before 1990.[16]

Although most public concern in Norway has concentrated on the loss of valuable fish resources, it should be emphasised that the acidification of such a large area of lakeland implies severe effects on whole ecosystems with far-reaching consequences. The acid input has completely altered the structure of many aquatic systems; appreciable amounts of poisonous metals have leached into groundwater; lakes have become overgrown with sphagnum mosses; many local species of insects, snails, crustaceans, mussels and plankton have been wiped out.[17]

The progress of acidification has been well documented in Norway. It was first recorded among the sensitive salmon populations, then among trout. Trout deaths began in small headwater lakes, and gradually spread downstream to include larger systems. Today, lakes in the southern areas which are classified as having sparse and decreasing populations are generally larger than those which are already barren.[18]

This must be particularly frustrating for the people living in the southern counties. They have seen the smaller, shallower lakes succumb to the acidifying process, and now they know that the remaining larger systems are also at risk. What is more, they can do nothing about it. Two scientists, Ivar Muniz and Helge Leivestad, reported to the SNSF project in 1980 about the depth of feeling they had encountered:

> We have been surveying area after area and seen the same story repeated over and over again, and have also felt the hopelessness and frustration of the people of these districts.
> What we are now facing is a major ecological disaster, and environmental degradation with a regional scope which is hard to grasp.[19]

Remedial measures

Ninety per cent of the airborne sulphur compounds falling to the ground in Norway comes from overseas, so any reduction of acidification depends upon a decrease of emissions from other countries. The Norwegians do not use an appreciable amount of coal. In 1977 they required all new oil-burning installations in polluted areas to use oil with a maximum of 1 per cent sulphur content, and are considering the extension of this policy to existing plants.[20] The Norwegian Bureau of Central Statistics claims that SO_2 output from 1980 to 1982 dropped from 140,000 tonnes to 110,000 tonnes per year, although this was partly caused by reduced industrial activity.[21]

Norway uses a limited liming programme for its fish farms, as a short-term remedial measure. As their Minister of Environment stated in August 1982, chemical treatment does not restore aquatic systems to their pre-acidification state; treatment of rivers and streams is difficult, if at all possible, during the period of spring snowmelt; and in the areas that get heavy rainfall, the water in lakes and streams is replenished frequently, so large-scale liming is necessary.[22]

Norway supported the proposal, at the June 1983 meeting of the United Nations Economic Commission for Europe, for an across-the-board 30 per cent reduction in SO_2 emissions.

Federal Republic of Germany

There is believed to be a large amount of continuing national research.

Effects

In 1982, 7.7 per cent of the forest area of West Germany was reported to be suffering damage (75 per cent light, 19 per cent medium, 6 per cent severe) from a wasting disease due to the consequences of air pollution.[23]

In 1983, a more thorough national investigation was launched, and the methods of data collection in each area standardised. Under the new system of analysis, the proportion of damaged trees leapt from 7.7 per cent to 34 per cent. In Bavaria, 46 per cent of trees are damaged; in Baden-Wurttemberg, which includes the Black Forest, 49 per cent of trees are suffering.

The survey revealed a disturbing increase in the number of trees suffering from medium or severe damage. In the space of 12 months, this jumped fivefold—to one tree in ten. Three out of four fir trees are damaged, with one in twelve dying. The director of Baden-Wurttemberg's forestry commission has warned that there is a real risk of whole forests being destroyed.[24]

The situation is now regarded as a national disaster, according the The Times (29 October 1983). It is probably difficult for people in the UK to understand either the extent of the problem or the German reaction to it. About 26 per cent of West Germany is forested, compared to 7 per cent of Britain. The area of trees suffering medium or severe damage—700,000 hectares (nearly 3,000 square miles) is equivalent to three-quarters of the total woodland in England.[25] The Germans are faced with the destruction of a part of their country which features prominently in their cultural heritage and national psyche. And the scale of the problem is enormous.

Brian Moynaham of Sunday Times visited West Germany in June 1983:

The Black Forest is dying.

'If pollution continues at the present rate, most of the fir and spruce trees in the Black Forest will be dead by the 1990s,' says Gerhard Weiser, agriculture and forestry minister for Baden-Wurttemberg. In the past two years, say ministry specialists, the number of healthy firs has dropped alarmingly from 66 per cent to only one per cent. In some areas, 94 per cent of spruce trees are also affected.

The early symptoms are easy to miss. The disease starts below the tops of the conifers. A process known as 'sub-top dying' turns the needles an attractive yellow-green. Growth is stunted, and the lower branches shed their needles. Lacking resistance to the cold, the bark splits in winter and bark beetles swarm in summer.

'Two years ago, there wasn't an unhealthy tree here,' says Karl-Viktor Gutzweiler, head forester of a 15,000-acre section of Black Forest near Baden-Baden. 'Now, half my firs are sick, really sick, and a third of my spruce are going.

'When you cut down a tree, you can tell from the rings that the disease started ten years ago. Until then, you got regular, well-spaced rings showing good growth. By the time the symptoms are visible, it is already too late.'[26]

The president of the German woodland owners' association claims that £13,000 million-worth of trees has already been lost;[27] over 500,000 hectares (2,000 square miles) of forest have been described as 'completely devastated areas', i.e. total loss of productivity is expected; and the decrease in forestry production will bring an estimated loss of 47,000 jobs in the German forestry and wood processing industry.[28]

There has been heated debate as to the causes of the death of Germany's trees. Some scientists have suggested ozone as the main culprit, others have put forward drought, beetles, acid mists, acidification of the soil, and bad forest management (this latter is normally suggested when there are no German foresters present, as the reaction to this tends to be rather explosive).

Professor Ulrich of Göttingen University originally alerted the Germans to the destruction of their forest land. His department's research has led to the belief that acid deposition builds up in the soil over a long time, with no observable effects until a damage threshold point is reached. During this stage, aluminium accumulation in the soil reduces the fine roots of trees, making them less able to absorb nutrients. The root structure changes, and the tree can sometimes be blown over by a heavy wind. Magnesium uptake by the tree is retarded, and the organism starts showing signs of magnesium deficiency—a yellowing and loss of needles. The tree suffers what Ulrich calls a 'decrease of vitality', and in this weakened state, is susceptible to attack by drought, beetles, ozone, acid mists, sometimes working together, sometimes separately.[29]

The theory—which enjoys more support than any other hypothesis for the widespread destruction of the German forests—points to ecosystem stress, rather than purely acidification, as the underlying cause of the collapse of the trees. The forest soils are not in a naturally balanced state but are continually being destabilised by acid deposition which is accentuated by the direct action of pollutants and other external agents on the tree leaves. Remedial measures would have to tackle both the driving force for destabilisation—acidification—and the external agents—ozone, acid mists, beetle attack. The first two agents can only be combated by a reduction of SO_2 and NO_x emissions.

Remedial measures
From a press release by the United Nations Environment Programme, 5 June 1983:

West Germany is to spend up to five *billion* dollars over the next ten years to fight the fast-growing international problem of acid rain.

This decision, personally announced by Chancellor Helmut Kohl, is the most important action yet taken by any country to combat the problem, which is turning between five and ten million square kilometres of Europe and North America acid ...

The German measures are particularly significant because the country has been one of the main producers of the pollution, which comes from the burning of fossil fuels ...

The measures are aimed at halving the amount of sulphur

dioxide emitted from power stations and large industrial factories by 1995. New pollution limits have been laid down and plants will either have to introduce controls to meet them or be closed down.

The chancellor estimates that the measures would cost the West German economy between two-and-a-half and five billion dollars over the next ten years. The country is also pressing the European Economic Community for new controls on pollution from car exhausts, the other main source of the trouble. [Emphasis added]

The legislation involves a 60 per cent reduction of SO_2 and a 40 per cent cut in NO_x emissions from large plant, essentially power stations, and has been used as the basis for the draft EEC regulations of 15 December 1983.[30] The Germans are also concerned to curb car-based pollution, and have legislated for lead-free petrol to be introduced on 1 January 1986,[31] after which new cars will eventually be fitted with tri-catalytic converters to reduce emissions of NO_x, hydrocarbons, and carbon monoxide.

West Germany supported the UNECE call for a 30 per cent reduction in SO_2 emissions, and is the driving force behind EEC moves in the same area.

United Kingdom

Research

In 1983, the Department of Environment spent £600,000 on acid rain research, which will increase to £1 million in 1984.[32] A good deal of work is done by the Institute of Terrestrial Ecology, which has established a monitoring network based in Scotland, and is studying the effects of acidification on forests, lakes, and the environment in general. Individual institutes have also run useful projects, but there is no national monitoring or evaluation system covering the whole of the UK.

The Central Electricity Generating Board (CEGB) is probably the largest researcher into acidity in the UK and has, according to one of its scientists, published 150 papers on the subject so far.[33] The extent and quality of its research programme is is difficult to establish. In 1982 the organisation issued a press statement claiming that it had spent £2 million on acid rain research in the financial year 1981−82,[34] but when asked to provide a detailed breakdown of this figure in 1984, was unable to do so. 'We are still feeding our research information into the computer', I was told by the press office, 'and we may be able to give you this information at a later date.'

The CEGB and the National Coal Board (NCB) have just co-sponsored a £5 million research project into the acidification of the lakes of Scandinavia, organised by the Royal Society and set to run until 1988. The study may be useful, though this is still the subject of some debate. Environmentalists in the UK want to know why the money is not being put into a survey of British damage; and some Norwegian scientists believe that the new project is merely 'rediscovering the wheel'. But the scheme's overall value has been discredited by the CEGB's announcement that they will not clean any of their SO_2 emission sources until the research is completed. This was perceived as a manipulation of scientific research in order to buy time (see page 75).

Effects

General According to a report published in January 1984,[35] prepared at the request of the Department of Environment, the rain in Britain is already acid, ranging from pH 4.7−4.4 in the west to 4.3−4.1 in the east. The western part of the country receives more rainfall than the east, however, so in total it suffers a greater input of acidity. The areas with the largest predicted acidic deposition are parts of Cumbria, the west Central Highlands and the southern uplands of Scotland; they currently receive acidic loads similar to those causing widespread devastation in Scandinavia and North America.

The monitoring systems used to work out these observations stretch through Scotland, northern England, the eastern Midlands and a small area on the south coast. The systems do not cover 90 per cent of Wales, any of Northern Ireland, or two-thirds of England, about which there is 'insufficient data' from which to draw conclusions. The observation stations are generally in areas of low rainfall, and not situated at high altitudes where the most acidic deposition occurs. They are also screened as far as possible from the pollutants created in towns. For these reasons it would not be surprising if a fully-developed monitoring system established the average rainfall over Britain to be *even more* acid than it is currently assumed to be.

Friends of the Earth analysed the UK Review Group's report, and pointed out that it says:

- Britain's rain is commonly 100−150 times more acid than 'normal' rain, and in one case 600 times more so.
- Britain 'exports' 76 per cent of its sulphur emissions abroad.
- Other countries account for only 11 per cent of all acidity in British rainfall.
- The report shows acidity occurring in short bursts. For example, 30 per cent of the acidity of rain falling at Goonhilly in Cheshire fell in just five days, with 50 per cent falling in ten days. This concentrated injection of acidity could have serious consequences for the environment.
- The report concludes that for the UK in recent years, evidence shows 'increasing acidity' in rainfall.[36]

Scotland *The Observer* recently published the results of a 1979 Norwegian survey of the Galloway and Loch Ard areas in Scotland. The researchers found that almost all of the 87 lochs and 47 streams surveyed were being turned acid by pollution. Each of the 26 medium and large-sized lakes in an area of 52,000 hectares (200 square miles) of west Galloway was so acid that life in the water was dying; at least two lochs in the area had lost all their fish, and in another, fish life was declining. Ten other lochs were approaching danger level; 30 of the 40 streams sampled in the Galloway area were acid enough to kill water life and seven of the rest were in danger. Ten of the 15 lochs and streams in the Loch Ard area were lethally acidic, and two more were close to becoming so.[37]

Information on Scotland was updated in December 1983 when Dr Morrison of the Department for Agriculture and Fisheries for Scotland said that his organisation had now monitored seven lakes that contained no fish life, and eleven in which fish stocks were becoming scarce. There was documentary evidence that Loch Enoch in Galloway had fish in it in the 1940s; and that Loch Fleet, in the same area, contained fish life until 1972. Both lochs have since become barren. Dr Morrison gave evidence of the maximum pH levels from seven lochs in Galloway. The pH levels varied from 6.8−6.1 in an eight-year sampling period from 1953 to 1961. In 1983, the spread was 5.7−4.92. As Dr Morrison said, there appears to have been an increase in local acidity in recent years. He also stated that this increase was not due to natural causes.[38]

Other evidence from Scotland is not yet adequately collated. In November 1982, an acid snowmelt killed 10,000 rainbow trout valued at £5,000 in a Galloway fish farm.[39] There has been concern in Glasgow because the acidity of the drinking water there has been dissolving lead in the water pipes. A 1979 Glasgow study revealed that one in ten children had higher blood levels of lead than the maximum regarded as acceptable by the EEC; and 50,000 people in Glasgow were reported to be suffering from low-level lead poisoning. Glasgow's drinking water is now limed to neutralise its acidity. Ayr's water has been limed since 1981.[40] How much these different factors are connected with acid rainfall is a point that needs to be established.

Wales In April 1983 the Scientific Director of the Welsh Water Authority issued a report about the death of fish in mid-Wales:

Surveys of a number of streams in the Upper Tywi Catchment and several lakes in mid-Wales suggest that many of the upland streams, rivers and lakes draining afforested catchments in Dyfed and Gwynedd cannot now support natural fish populations and have depleted populations of aquatic plants and animals. Attempts to restock the River Tywi above Llyn Brianne with trout and salmon have proved unsuccessful and fish survival tests have shown that native brown trout cannot survive the combined effects of acidity and elevated aluminium concentrations found in water draining from conifer forests in the area.

Lakes in afforested catchments such as Berwyn and Blaenmelindwr have been recognised as marginal fisheries for some years and they have recently been operated as 'put and take' fisheries using American Brook Char since these are more tolerant of acidic conditions than natural species. This is not, however, a satisfactory solution to the problem since Berwyn is now too acidic to support Brook Char and although some stocked fish are surviving at Blaenmelindwr, the genetic implications of introducing exotic species are causing concern. Similarly, stocking marginal fisheries with hatchery-reared fish of low acid tolerance may reduce the resistance of natural stocks to further acidification.

...In addition to the implications for fisheries described above, acidic waters can also cause water supply problems. Failure to neutralise acidic waters can result in excessive corrosion of mains, with, in some areas, unacceptably high lead levels due to plumbosolvency [dissolving of lead] and consumer complaints due to coloured water. If upland waters are in fact becoming more acidic then additional treatment costs appear to be inevitable.[41]

In September 1983 *The Observer* reported that an alarming draft report was being circulated confidentially by water authority scientists. The paper said, wrote Geoffrey Lean, that

all the fish and most of the other life in important mountain waters in Dyfed have been killed. The Welsh Water Authority is seriously worried that the whole of upland Wales from Snowdonia to Camarthen could be affected ...

Water authority scientists have also found acidic streams and lakes east of Aberystwyth, and they fear that the whole 100-mile long mountain range from the North Welsh coast to Dyfed may be in similar trouble.[42]

The rainfall in mid- and west Wales is among the least polluted in northern Europe.[43] There are, however, occasional outbreaks of strongly acid rain which probably result from easterly winds transporting acidic emissions from urban areas of England. The geology of the area, thin soils on hard rock, is unable to buffer these occasional high-input acid attacks.

The Farmers' Union of Wales is seriously concerned about the acidification of the border areas with England. 'The situation is very serious,' said Emlyn Thomas, Brecon and Radnor secretary of the FUW in July 1983. 'We are pointing out the dangers and urging careful monitoring. In addition we are urging that any necessary control measures are carried out as speedily as possible.'[44]

It is worth noting in passing that this growing acidification was discovered by accident. Water scientists were monitoring the Llyn Brianne reservoir which, after its construction, seemed to be having an adverse effect on the availability of local fish. It was only when the cause of this dearth was discovered — acid water running into the reservoir — that the water authority began to suspect there was a regional rather than merely local problem here. Had there not been a need to check the effect of the new reservoir on the environment, it is possible that the growing acidification of Wales would not have been discovered until a later date.[45] Having said that, it should be pointed out that the Welsh Water Authority is now conducting extensive research into acidification of streams and lakes, and in October 1983 set up a system for monitoring acidity of rainfall throughout Wales. WWA scientists have also produced some of the most up-to-date reports in the UK on the effects of coniferous forests and acid deposition on water quality.

England Little is known about the effects or spread of acidification in England; monitoring stations cover only about one-third of the country. The three sites in the south-east used by the UK Review Group on Acid Rain indicated strong average acidity of rain (pH 4.1); this, according to the report, is the most severe in the UK.

Some places in the UK, near industrial sources, have experienced acid deposition for many years. Rain on the outskirts of Manchester and Leeds has contained strong acids for more than a century. Some of the most acid moorland pools and lakes in Britain occur on the southern Pennines near Huddersfield and Sheffield, where pH values of 3.3 to 3.5 have been found in surface water. Several moorland waters in Northumbria have a pH of 3.9 to 4.0, and have had for the last twenty years.

In south Cumbria, the pH of precipitation has varied from 3.5 to 6.8, with an average of 4.4 since 1959. When the wind blows from the Atlantic Ocean, the rainfall is less acidic; when it comes from the south and east, the rainfall is more acidic.[46] The Lake District is an area particularly at risk from acidification because of the combination of its geological make-up and a high acidic input. *Trout and Salmon* has drawn attention to the loss of an entire run of salmon and sea trout in July 1980 in the River Esk, Cumbria, the cause of which is commonly assumed to be acidification. The publication also noted the increasing lack of salmon parr in the rivers Spey and Ribble. The North-West Water Authority is currently conducting its own study of high altitude streams.

The authorities in England are coming increasingly under attack for their lack of research into acid pollution. Professor Last, one of the leading British authorities on the phenomenon, wrote in May 1983 that British knowledge of acid effects in rural areas is 'woefully inadequate'.[47] He was perhaps unconsciously echoing the earlier testimony of the Natural Environment Research Council to the Royal Commission on the Environment (1976):

Air Monitoring Networks: Despite arguments to the contrary, most government-sponsored networks are based on obsolescent technology directed towards monitoring pollutants of concern to human health, and have little immediate relevance to rural Britain and effects upon crops, trees, natural vegetation and soils.

Vulnerable areas in England include the woodlands of Cambridgeshire and Essex, Charnwood Forest in Leicestershire, the New Forest, the Lake District, and the woods and crops around London.[48] There is no way of knowing, however, whether crops, lakes, or trees are being damaged in these susceptible areas because the research is not being done. As one environmental body (SERA) has put it: 'No money equals no research equals no problem.'

Northern Ireland It has not been possible to find any monitoring or evaluation work related to acid deposition in Northern Ireland.

Remedial measures

For the last 150 years, the UK has been the largest producer of airborne sulphur dioxide in Western Europe, and it continues to be so. In Europe as a whole, its SO_2 production is second only to that of the USSR which, unlike the UK, is a net importer of SO_2 and has also agreed in principle to a 30 per cent reduction in transboundary emissions by 1995.

The UK's total sulphur dioxide output has dropped from a total of 5.65 million tonnes in 1972 to 4.04 million in 1982 − a 28 per cent reduction in ten years, achieved partly by increasing use of natural gas, partly by a switch to nuclear fuel, but mainly because of the increasing price of oil and Britain's industrial recession. The two main falls in SO_2 output − 450,000 tonnes in 1974, and 710,000 tonnes in 1980 − occurred after the oil price rises of 1973 and 1979, when the government, for economic reasons, switched from oil consumption to cheaper (and coincidentally cleaner) fuels, and at the same time conserving energy use.

The 1980 drop in emissions was also a reflection of the severity of the slump in the UK after the onset of the new government's economic policies: the reason for the reduction in industrial emissions was the reduction in industry. Energy consumption in Britain fell by 7.8 percent in 1980 and by a further 5 per cent in 1981:[49] a 13 per cent fall in consumption goes a long way towards explaining the 20 per cent drop in UK SO_2 output between 1979 and 1981, with the increased price of oil taking care of another few percentage points. The figures, which have been taken from the Commission on Energy and the Environment Report, 1981, and updated by the Department of Environment, make interesting reading, as shown in Table 4.1. Although the national output fluctuates, it moves up and down as an indicator of economic performance: the figure stays around the 5 million mark through the middle and late seventies, and then plummets in 1980. The power station figures, by contrast, have stayed remarkably steady. The question arises: what will happen if and when the economic recovery occurs? The CEGB have already answered this. In a letter to *The Times* on 2 April 1983, the Secretary of the CEGB acknowledged that, if the UK increases its Gross Domestic Product by 2.6 per cent per annum until the year 2000 — a figure which the government is currently claiming to be perfectly feasible — then emissions from power stations will increase by 30 per cent before the end of the century. This would produce an emission figure of 3.45 million tonnes of SO_2 from the power industry, and maybe 4.9 million tonnes from the UK; and this at a time when most other countries in Europe are trying to cut back their airborne pollutants. It is important to realise that the UK's satisfaction over its reduced sulphur dioxide emission figures is predicated on continued economic recession. If, as the government promises will happen, the economy picks up, then our SO_2 emission figures will leap towards their earlier peaks.

Table 4.1 UK national and power station SO_2 emissions (million tonnes)

Year	UK output	Power station output	Percentage SO_2 from power stations
1972	5.65	2.87	51
1973	5.81	3.02	52
1974	5.36	2.78	52
1975	5.17	2.82	54
1976	5.00	2.69	54
1977	5.00	2.74	55
1978	5.02	2.81	56
1979	5.38	3.1	58
1980	4.67	2.87	61
1981	4.29	2.71	63
1982	4.04	2.65	65

While we are on the subject, it is worth comparing the power station emission figures for 1972 — 2.87 million tonnes — with the 1980 output — 2.87 million tonnes. Even in the middle of a recession our power sector's record for reducing its emissions is abysmal. And if abysmal is the correct word to describe the electricity industry's attitude towards SO_2 pollution, there is another to describe the UK's remedial policies for curbing SO_2 emissions. The word is: non-existent.

The UK did not support the UNECE proposal for a 30 per cent reduction in European SO_2 emissions.

Switzerland

The pH of precipitation in Switzerland has recently been established as an average 4.3. A 1983 report by Lausanne Federal Polytechnic[50] states that 'the acidification of mountain lakes in southern Switzerland and damage to forests indicate serious dangers for the environment'. The report also states that 'in Germany, the Ministry of Agriculture estimates that 8 per cent of its forests has been damaged or is threatened by pollution … the situation in Switzerland is no different to that of Germany as far as pollution is concerned.'

Switzerland has good reason to be anxious about any threat to its forests. During the nineteenth century the country was shaken by a number of avalanches, the cause of which was discovered to be the indiscriminate felling of trees on mountainsides. Swiss mountain villages are traditionally built below protective forest layers, and when these are removed the townships are vulnerable to landslips from further up the peak. Nowadays, there are strict federal laws governing tree harvesting and the amount of wood cut per year. If acidification starts killing the trees on a large scale then the consequences could involve loss of life.

Switzerland has announced that it will move to lead-free petrol in 1986 — the first step towards cleaning NO_x from traffic emissions. It also supported the June 1983 proposal at the UNECE for a European reduction of SO_2 emissions by 30 per cent.

Austria

The effects of acidification began to become increasingly more apparent in 1983. According to *The Times* of 19 November 1983, 'Acid rain has increased dramatically in the last year, destroying 296,000 acres of Austrian forest. In some southern valleys, a third of the trees are dying. To combat this, the Austrian government will subsidise lead-free petrol and reduce by half the level of toxic chemicals in heating oil.'

Austria supported the June 1983 proposal at UNECE for a European reduction of SO_2 emissions by 30 per cent.

Denmark

Most Danish lakes, especially the larger ones, are typical bicarbonate systems with strong buffering capacities. There is a tendency towards acidification in some soft-water systems, especially the Lobelia lakes, which are becoming more and more rare now in Denmark.[51]

Denmark supported the 1983 UNECE proposal for a 30 per cent reduction in SO_2 emissions.

The Netherlands

In eastern Holland, premature ageing of needles in Scots pine and Douglas fir has been found over a wide area. In May 1983 acute damage to pine trees was found in places.[52] Later in the year, the State Forestry Service looked at 1,500 stands of trees and found that virtually all the conifer woodlands in the country were 'lightly damaged' by acidification. Scots pine appears to be the most severely affected.[53]

In 1965, the Netherlands produced 900,000 tonnes of SO_2 per year. A switch to natural gas reduced this total to 450,000 over the next ten years. When it became apparent in 1979 that the country was going to depend more on coal for energy production, the Dutch government announced a national SO_2 output ceiling of 500,000 tonnes per year. New coal-burning installations now have to obey SO_2 emission restrictions.[54]

The Dutch Environment Minister said in October 1983 that current emissions of SO_2 and NO_x in Europe should be reduced by a factor of three or four. As a first step towards this 'multilateral' aim, a 5 per cent reduction in SO_2 was proposed to the Dutch parliament in late 1983, and a new ceiling for NO_x emissions.[55] This was converted into a 30 per cent reduction pledge at the meeting of the 'Club of Thirty' in March 1984.[56]

France

The French have been monitoring their eastern forests since realising the dangers to the forested areas in Germany. In the summer of 1983 they first noticed unusual signs of tree collapse: at the end of 1983, these signs, or at least the official reports of them, multiplied. The French Ministry of the Environment now estimate that several thousand hectares of woodland have been affected, mainly hillsides in Lorraine and Alsace. Spruce and pine show most damage. The Ministry also note that precipitation in the affected areas has become three or four times more acid since 1974.[57]

On 22 February 1984, the French government announced that it would reduce its sulphur dioxide emissions by 50 per cent between 1980 and 1990, and would try to obtain agreements from other countries to develop coherent pollution reduction programmes.[58]

No announcement was made as to the costs of these measures, but estimates by the OECD in 1978 suggested that the price of the French programme will be about £2 billion.[59]

Italy

Some damage to pine trees has been reported in Italy,[60] but the major effects of air pollution are to historic monuments (see page 36). Of the West European countries that are currently refusing to cut industrial emissions, Italy is the second largest SO_2 producer.

Belgium

Belgium holds the dubious distinction of having a higher rate of sulphur dioxide emission per square mile than any other country in Europe. Research into acidification is still not very far advanced, although the pH of rainfall in the Campine Fens has been established as 3.8, and moorland pools in the area have become 100 — 1,000 times more acid over the last forty years.[61] Belgium soil appears to have good buffering qualities and widespread damage has not yet been recorded.

Yugoslavia

In the Yugoslavian part of the Erzgebirge, nearly 500,000 hectares (1,900 square miles) of woodland have perished since 1970 and 52,000 hectares (200 square miles) of woodland are seriously affected.[62]

Greece

In the first few days of 1984, the poisonous cloud of pollution hanging over Athens forced the Greek government to order a 30 per cent reduction in industrial production, reduce central heating in public buildings, and ban some vehicles from the capital.

As the smog persisted the authorities declared a zone of 9,000 hectares (36 square miles) in the city in which cars and taxis were allowed to travel only on alternate days. People with lung problems were warned to stay home until the pollution ended. The new measures were the strictest ever imposed in the city.[63]

Eastern Europe

Information on sulphur dioxide output and the extent of acidic damage in Eastern Europe is hard to quantify: the former has to be guesstimated and the latter taken from international sources rather than national surveys. The region is a large producer of SO_2 (see page 54), and in order to stop the spread of acidification it is essential that Czechoslovakia, the GDR, Hungary, Poland, Rumania and the USSR reduce their emissions. The first five countries produce per capita significantly larger amounts of SO_2 than most West European countries, and their spillover pollution has damaging effects on West German forests and Scandinavian lakes.

This SO_2 contribution presents a dilemma to West European environmentalists. Western European industrial lobbyists frequently point out that the sulphur dioxide produced in Eastern Europe forms two-thirds of the European total; why, they then ask, is the acid rain campaign concentrating against the West rather than the East?

The answer lies in the selective presentation of statistics. The East European output of 39 million tonnes of SO_2 per year includes 25 million produced by the entire European and Asiatic land mass of the USSR. As there are 8,000 kilometres (5,000 miles) between the east and west frontiers of the USSR, and acid rain is commonly assumed to travel a maximum of 2,000 kilometres (1,250 miles), it seems unlikely that a power station in Okhotsk is likely to have much effect on Western Europe. Even if all the Asiatic lands of the USSR are included, however, that country is still a net importer of sulphur dioxide emissions because of its geographical position; and there is a rough balance between SO_2 imports and exports in the six countries mentioned above plus Bulgaria.

This is not to belittle the part played by the polluters of Eastern Europe; the Green Party in West Germany is rightly concerned about the contribution of East Germany towards the acidification of the West German forests. On the other hand, it is important that the electricity industries of the West do not hijack this concern into making the East the scapegoat for all, or even the greater part of, acidic pollution in Western Europe. In Norway, for example, twice as much acidity arrives from Western European pollution than from the East.

Poland
More than 388,500 hectares (1,500 square miles) of woodland have been damaged by acidification. The heavily wooded

mountains between Poland and Czechoslovakia are seriously affected.[64] Forest researchers in Katowice, near Krakow, say that fir trees are dead or dying on nearly 180,000 hectares of land and that spruce trees in areas around Rybnik and Czestochowa are completely gone. Environmental scientists warn that by 1990 as many as 3 million hectares of forest (11,000 square miles) may be lost if Poland proceeds with its present industrialisation plans.[65] For damage to buildings in Poland see pages 32−33.

Czechoslovakia

In the Erz mountains of western Czechoslovakia, nearly 103,000 hectares (400 square miles) of forests are already dead, where less than ten years ago healthy trees covered the mountainsides.[66]

Pollution in Czechoslovakia is reaching crisis point, according to a 28-page report commissioned by the Czechoslovak Academy of Science, smuggled into Austria and released by Charter 77. Already 37 per cent of Czech forests−480,000 hectares (1.2 million acres)−is either irreparably spoiled or dead. Pollution in the air and water is 10 times the world average, and if no action is taken whole species of plants and animals will vanish, as will 60 per cent of forests, by the year 2000. Four or five major cities are already supplied with drinking water that is a health hazard; within sixteeen years, parts of the country could be unfit for human habitation; tests in Bohemia show that children have 20 per cent more red blood cells and slower bone growth than in less polluted regions. Life expectancy among adults in Bohemia is three years lower than the national average. (Report in *Toronto Globe and Star*, 12 January 1984.)

German Democratic Republic

The devastated woodlands in the Erzgebirge have upset the water economy on the northern slopes and this is having very serious consequences for the lower-lying agricultural areas. Some 12 per cent of East Germany's forests are believed to be affected.[67] Trout stocks have either vanished or suffered marked reductions compared with thirty years ago. Lake acidification has been detected in Thuringer Wald and Erzgebirge; and pH values as low as 3.9 have been recorded in some places.[68]

USSR

There is an unconfirmed report, from a Swedish journalist who got the information from Soviet forestry commission officials, that 360,000 hectares (900,000 acres) of Soviet forest have so far been acidified.[69]

A senior Soviet official stated in February 1984 that damage caused by acid rain to agricultural production was causing serious concern in the USSR. The deputy director of the Institute of Applied Geophysics in Moscow said that estimates put the yearly damage to Soviet agriculture as high as $500 million (£330 million). The pollution affects up to 15 per cent of the harvest in 41 million hectares (160,000 square miles) west of Moscow.[70]

Pravda recently revealed that vast areas of forest are dying from air pollution near the car-manufacturing city of Togliatti, about 1,300 kilometres east of Moscow. According to the report, nearby forests along the Volga River may soon resemble a wasteland.

Rumania

56,000 hectares of Rumania's 6.3 million hectares of forests are damaged by industrial emissions.

Canada

Research

There are currently 350 projects researching into the acidification of the environment, funded by federal and provincial authorities, and costing about $25 million per year (£17 million). About 300 full-time scientists, technicians, and support staff are employed; 40 per cent of the work deals with atmospheric processes, 40 per cent with aquatic effects, and 20 per cent terrestrial damage (mainly forestry).[71]

Effects

The effects of acid rain in North America are similar to those experienced in northern Europe. To date, no forest damage has been defined in Canada, but a widespread threat has been identified in terms of acidification of surface waters. The lakes in eastern Canada, as in Norway and Sweden, lie on granitic or gneissic rock with little buffering capacity. An acid-sensitive area comprising 2.5 million square kilometres (a million square miles) takes in Ontario and Quebec, and contains over a million lakes.

The eastern half of North America has a substantial load of sulphur and nitrogen oxides emissions, with winds in both winter and summer moving towards the north-east. Although the number of dead lakes in Canada is still relatively low – in the order of several hundred – it is the frightening extent of possible destruction that is causing concern: 2.5 million square kilometres of sensitive territory is at risk. A survey of 4,500 lakes in eastern Canada has indicated a significantly lower buffering capacity for lakes in the danger region, compared with waters of similar nature outside the area. Surface waters are under stress, and are changing because of the combination of the acidic input and their geological characteristics.[72]

In Ontario, 20,000 lakes are in imminent danger of acidifying to the point of losing fish life, with another 28,000 directly under threat.[73] In Nova Scotia, eight rivers are now believed to be completely empty of salmon, thirteen are becoming incapable of supporting salmon fry, and another nine are acidifying to the same dangerous level.[74] About 30 per cent of Nova Scotia's salmon industry is in danger of being wiped out.

In Quebec and Labrador, the precise degree of acidification of lakes is still being documented, but research suggests that wind patterns and lake chemistry make it likely that future acid rain effects could be severe. Nearly all of Quebec's surface waters are believed to be highly sensitive to acidification. Monitoring of Laurentides Park, with its 2,500 lakes, has shown that rainfall in the area is consistently acidic.[75]

The Canadians are extremely worried because, to quote H.C. Martin, Senior Adviser to the Long Range Transport of Air

Pollution Liaison Office of Environment Canada, 'Over the last few years we have observed a continuous degradation of the biological health of a large segment of the environment.' Although, as mentioned earlier, Canadian scientists have not monitored any loss of forest productivity to date, they are concerned by the West German experience. The economic consequences of forest loss in Canada would be considerable, as the forestry industry and associated businesses employ 10 per cent of the Canadian workforce and, in the east, contribute $11 billion to the Canadian Gross National Product every year.[76]

Other damage effects are still being discovered. According to Martin Weaver, technical director of Heritage Canada, acidity causes about $1 billion-worth of damage to buildings in Canada each year. Human lungs, according to a report cited in an article in the *Ottawa Citizen*, can be eroded by acid-laden moisture in the air. One American scientist has estimated that as many as 5,000 Canadians die annually from acid-related bronchial problems. Canadian federal health officials say this figure is too high, and the department is trying to get a more precise reading.[77]

It is worth repeating the speech made by the Canadian Minister of the Environment in April 1983:

> The Canadian economy is dependent on our natural resource base. This natural resource base is threatened by acid rain. The gross economic activity generated by fishing in eastern Canada exceeds $1 billion a year. Tourism revenues are more than $10 billion and forestry produces more than $15 billion a year. Altogether, we are talking about a substantial portion (about 8 per cent) of our Gross National Product ... at risk because of acid rain.

Remedial policies

Over 31 million tons of sulphur dioxide and about 22 million tonnes of nitrogen oxides were emitted into the North American atmosphere during 1979. Of this total, Canadian sources accounted for 5 million tonnes of SO_2 and 2 million tonnes of NO_x.

More than half (2.7 million tonnes) of this SO_2 was produced by 10 plants, mainly smelters and power stations. One plant in particular, the Inco copper and nickel smelter, emitted 866,000 tonnes — a phenomenal amount, roughly equivalent to the total produced by a small country such as Belgium. In the mid-sixties, this plant emitted an astounding 1,800,000 tonnes of sulphur dioxide a year.[78] This was reduced to its present level by local legislation, and in January 1984 cut to 711,000 tons. Planned new burner technology will further reduce output to 456,000 tonnes a year. All in all, the Canadians legislated to reduce their SO_2 emissions by 25 per cent in eastern Canada from 1980 to 1990 and at the Ottawa meeting of the Club of Thirty on 20 March 1984, this figure was increased to a 50 per cent reduction by 1994[79].

The Canadians have a problem that Norway and Sweden would understand: spillover pollution from another country. Four million tonnes of SO_2, and an unknown quantity of NO_x, drift across the border from the USA into Canada every year.[80] A 100 per cent reduction of emissions in Canada would not adequately protect the environment because approximately half of the sulphur dioxide in the Canadian airways comes from the States. Canada cannot hope to protect its lakes without a reduction in pollution from the USA. With this point in mind, the Canadians wanted a 50 per cent reduction of emissions by both countries in the eastern half of the continent to become effective by 1990.[81] The Americans are currently considering this proposal without much enthusiasm: in February 1984 the USA announced that no action was to be taken for the moment.[82]

United States

Research

In 1980, the US Congress passed the Acid Precipitation Act, which instructed the Environmental Protection Agency, in co-operation with the other members of the Inter-agency Task Force on Acid Precipitation, to identify the causes, sources and effects of acid precipitation. A 10-year project was set up, funded to the tune of $18 million (£12 million) in 1982 and $22 million (£15 million) in 1983.[83] Roughly 50 per cent of this is spent on monitoring and tracking emissions, with 40 per cent devoted to estimating effects. There are a number of other projects, funded both privately and by state authorities.

Effects
One of the problems of documenting forest damage is the lack of historical records. If a wooded area *looks* healthy, most people are not particularly interested in counting and measuring the number of trees it contains. It is only after the vegetation starts dying that efficient surveys are started.

An exception to this state of ignorance, however, is the Camel's Hump Mountain in Vermont. In the mid-1960s a graduate student called Tom Siccama made a thorough study of the mountain's vegetation, climate and soils. He also counted and measured the trees, which makes his study an invaluable data base for today.

In an article in *Natural History,* November 1982, H.W. Vogelmann, Professor of Botany and Chairman of the Department of Botany at the University of Vermont, described the current situation at Camel's Hump. The mountain has a high-elevation spruce forest regularly swept by rainfall twice as acidic as normal rain, and also by fogs 100 times more so. Situated to the north-east (and in the plume-tracks) of Ohio and Pennsylvania — the two largest SO_2 emitters in the USA — Camel's Hump can be regarded as a high-risk area for acidification. Vogelmann's observations appear to bear this out:

Using the Siccama data for comparison, researchers at the University of Vermont have been able to document that nearly 50 per cent of the spruces in the Camel's Hump forest have died since 1965. The density, basal area (a measure of the amount of standing wood) and seedling reproduction also declined about 50 per cent …

Today the red spruces are dead or dying and some firs look sick. Grey skeletons of trees, their branches devoid of needles, are everywhere in the forest. Trees young and old are dead, and most of those still alive bear brown needles and have unhealthy looking crowns. Craggy tops of dead giant spruces are silhouetted against the sky. The brittle tree tops break off, leaving only a jagged lower trunk with a few straggly branches. Strong mountain winds overthrow many dead trees, tipping upward their shallow root systems along with chunks of the forest floor. As more and more trees die and are blown down, the survivors have less protection from the wind, and even they are toppled over. The forest looks as if it has been struck by a hurricane ...

Spruces are succumbing throughout the northern Green Mountains, especially on the windward slopes at high elevations. Dead and crown-damaged trees are common in the Adirondack Mountains in New York, in the White Mountains in New Hampshire ... and in the Appalachians as far west as Virginia. It is a disaster that, in a few short years, has dramatically changed the appearance of high mountains.[84]

Vogemann implicates acid rain as the cause of this destruction, but points out that the evidence is still circumstantial. Interviewed by *Natural History* half a year later, however, he states:

Today, I would make even stronger statements, because our findings over the last six months have deepened the evidence linking acid rain and other pollutants to forest damages. We're getting very close to the point where the evidence can be characterised as more than 'circumstantial'.[85]

There is no doubt about surface water acidification. At the beginning of the eighties, Ellis Cowling, Chairman of the National Atmospheric Deposition Programme in the USA, wrote:

As of 1980, hundreds of lakes in the Adirondack region of New York State .... were showing acid stress in the form of diminished populations. Lakes and streams in other regions of the United States and Canada also are vulnerable to stress by acid precipitation. These regions include northern

Minnesota, Wisconsin, and Michigan, parts of southern Appalachia and Florida, large parts of Washington, Oregon, California, and Idaho and large parts of the Canadian maritime province.[86]

In 1982, the US National Wildlife Federation stated that 'acid rain may be slowly poisoning 100,000 miles of streams and 20,000 lakes'. It went on to say that at least 212 Adirondack lakes were fishless, 180 former brook trout ponds in the same area no longer supported fish, and in Maine, 1,368 lakes had experienced an eight-fold increase in summer acidity from the 1920s to the 1970s. Nine out of eleven lakes sampled in northern Florida were acidified, and overall:

38 − 81 percent (depending on the precise area measured) of the lakes and streams tested in the eastern US, and 13 − 48 per cent of midwestern lakes, fall in the categories of 'moderately sensitive to acid rain' to 'already acidified'.[87]

As for architectural corrosion, the Canadians have pointed out that the President's Council for Environmental Quality estimated in 1979 that the annual cost to the USA in architectural damage due to acidification was upwards of $2 billion.[88] The Canadian Minister of the Environment, John Roberts, has also drawn attention to Senator Mitchell's evidence to the US Committee on Energy and Natural Resources in August 1982:

The economic impact of acid precipitation on existing economic activities in the east is $5 billion annually. It is also estimated that damages to crops and fisheries from acid precipitation may total from $15 − 25 billion by the end of this century.[89]

Remedial measures

In theory, the USA has moved further on the control of acidic pollutants than many other countries. In 1970, the Clean Air Act was passed with the intention of controlling pollution at its source, by requiring states to reduce SO_2 and NO_x levels to nationally-set air quality standards. Each state had to prepare, implement, maintain, and enforce a control plan which is supervised by the Environmental Protection Agency (EPA). All new plants built after 1970 had to obey tougher new pollution standards—although the way in which they did this was to be at their own discretion.[90]

In 1975, the USA introduced lead-free petrol, and legislated that all new cars were to install catalytic filters in order to reduce NO_x, hydrocarbon, and carbon monoxide emissions.[91]

In terms of overall air pollution, the new measures have not been successful. Industrial response to the 1970 Act was to build tall chimneys to disperse the SO_2 and NO_x, thus satisfying the new regulations (which had no control over the long-distance pollutants sulphate and nitrate). Certain high-polluting states, such as Ohio, resisted the implementation of viable standards for at least five years and failed to prosecute offending plants when the new standards were forced on them by the federal authorities.

After the 'tall chimneys' loophole was plugged in 1977, some new plants were built with scrubbers to clean the emissions—in 1982, there were 87 scrubbers in place, with 22 in construction—but the problem is that plants built before 1970 are allowed to continue as before. With an expected plant life of up to 40 years, it is calculated that in the year 2000, 70 per cent of all SO_2 emissions in the USA will come from emission sources built before 1970. The EPA itself predicts that SO_2 emissions will not begin to decrease until the year 2000, give or take the odd economic slump.[92] This continuance of pollution should be considered in conjunction with a 1978 comment from the International Joint Commission, a bilateral body primarily known for its patient efforts in reducing the Great Lakes water pollution: 'Unless emissions are reduced, widespread, irreversible harm to ecosystems susceptible to acid rain will occur within 10 to 15 years.'[93]

The North Americans are profligate users of energy, consuming per head of population more power, and emitting more sulphur, than any other people in the world. The reductions they are making are too little, too late. Thus the introduction of filters on car exhausts, while reducing NO_x and ozone concentrations in Los Angeles county, are not producing similar declines nationwide.[94] The reason for this, presumably, is the lack of adequate control of stationary sources.

In 1982, legislative proposals were put forward to reduce SO_2 emissions in the 31 states east of the Mississippi by 35 per cent over the next ten years, at a projected increase in electricity costs of $4-8$ per cent.[95] For the next 12 months the Reagan administration, conscious of the forthcoming presidential election and worried about acid rain becoming an issue, flip-flopped over this and other desulphurisation proposals. According to one source in the EPA, a desulphurisation announcement was planned for an EEC acid rain meeting in September 1983, but industrial pressure eventually won the day.[96] It now looks as if the Republicans are going to try and bluff it out. If Reagan wins the 1984 election, the prospects for the environment look gloomy.

Other countries

The Arctic

This, one of the world's last pristine wilderness areas, has become a victim of acid precipitation. Investigations by Norwegian scientists have revealed concentrations of sulphur dioxide comparable to levels recorded in southern Norway. The result has been the creation of a phenomenon known as 'Arctic haze', which restricts horizontal visibility to $3-8$ kilometres, and probably covers an area almost equal in size to the North American continent. The effects of this haze on the environment are unknown. The pollution, which covers the Polar cap throughout the winter months and begins to clear up in April, is thought to absorb a substantial amount of solar energy during the spring months. Some scientists fear this absorption could result in a heating of the Arctic atmosphere, and consequently cause a change in the global climate.[97]

Brazil

Strong evidence of acidification exists in the low pH values of soil sampled in the eastern part of São Paolo State (3.7 to 4.7, compared with a state norm of about 5 to 7). This was found in

Cubatao and São José dos Campos, both within an hour's drive of the capital and its seaport Santos, where air pollution is an old problem.[98]

Cubatao is known among its 85,000 residents as 'the inferno' and 'the valley of death'. Its petrochemical and other industries pump some 750 tonnes of toxins—including sulphur dioxide, ammonia, carbon monoxide, and fluoride gas—into the air every day. This cocktail of pollutants means it is impossible to isolate SO_2 pollution from the rest.[99]

China

Acid rain in parts of China produces contamination as high as in the affected parts of Europe, according to the *Guangming Daily*, an official newspaper. Rain tested in 2,400 localities was found in 44.5 per cent of cases to be acid. The problem was largely restricted to the region south of the Yangtse River, and was especially serious in Canton, Suzhou, and Chongqing. Rice planted on 1,320 hectares (3,300 acres) near Chongqing had suddenly wilted and died.[100]

South Africa

Air pollution of a kind likely to cause acidification occurs in the eastern Transvaal Highveld of South Africa. Enormous oil refineries, coal-burning power stations and other industrial complexes clustered round Johannesburg and Witwatersrand will release about 2,220 tonnes of sulphur dioxide a day into the atmosphere by 1985. This, because of local meteorological conditions, will stay mainly within an area of about 3,000 square kilometres (1,200 square miles), a potential fallout of 221 tonnes of SO_2 per square kilometre from power stations alone. The equivalent figure for the Ruhr basin in Germany, a noted area of acidic pollution, is 260 tonnes per square kilometre from all sources.[101]

Australia

In Australia, acid rain is not considered a problem at the moment, but the government is establishing six monitoring stations around the country where air pollution might be expected to occur. One area, near Sydney, has so far yielded serious signs of acidification.[102]

Environmental Damage in Europe

Key

→ prevailing wind
🏭 country that emitted over 1 million tonnes SO₂ in 1982
🌲 significant forest damage
🐟 significant damage to lakes and streams
(30%) country that has announced at least 30% desulphurisation programme

Figure 4.1. From the map, it can be seen that the East European countries will have to be involved in a European desulphurisation programme. In Western Europe, five countries provide 75% of the area's SO₂. In terms of diplomatic initiatives, Spain and Italy are not prime targets because of their distance from Central Europe and because of direction of wind currents. Of the three remaining major polluters, the UK is the only one that has not yet started a desulphurisation programme.

Appendix A: Research into acid rain

Complaints about smoke and sulphurous fumes from the combustion of coal have been recorded since at least the thirteenth century. In 1306, Londoners were forbidden to burn sea-coal because of the type of smoke it produced. But the ban did not seem to stick: in 1661 John Evelyn condemned the 'hellish and dismal cloud of sea-coal which maketh the City of London resemble the suburbs of hell'; and in 1727, Hales noted that the air was 'full of acid and sulphureous particles'.

The term 'acid rain' was coined in 1872 by Robert Angus Smith, in a publication called *Air and Rain: the Beginnings of a Chemical Climatology*. In this book, he puts forward many of the ideas about acidification that have since been arrived at independently; but unfortunately the work did not have much of an impact in scientific circles until its recent revival in 1981. Sulphur dioxide pollution made a front-stage appearance in Britain in 1924, when a Manchester farmer sued Barton power station for compensation, on the grounds that the sulphur content of its flue-gases was damaging his crops. Although he lost the case when it went to the House of Lords, his situation aroused public comment and inquiry, and a good deal of anxiety. Coincidentally, this controversy occurred at the time that Battersea power station was being constructed, and as a response to various representations (including, it was rumoured, from Buckingham Palace, on the grounds that George V suffered from chest trouble) the Electricity Commission insisted that the new power station should install desulphurisation machinery. This explains why Battersea power station, for most of its working life, cleaned 80—90 per cent of the sulphur from its smoke.[1]

Much of the early research into acidification took place in Scandinavia, where for several decades in the twentieth century scientists were puzzled by the fact that fish in the lakes were dying. These deaths started slowly at first, in the early part of the century, but the trend increased in the 1940s, and has accelerated since then. Researchers were initially handicapped by the compartmental nature of science: soil scientists kept their eyes on the ground, and did not concern themselves with atmospheric input; atmospheric specialists were not interested in the effects of pollutants after they left the air; and the lake technicians — limnologists — did not keep track of what was happening in either of the other two disciplines. So, for some time, most people worked in their own separate areas, and missed the results that were being reported elsewhere.

One unfortunate example of this compartmentalisation occurred between 1955 − 65, when Gorham published a series of papers on acidification, which are now acknowledged to be the foundation of our present knowledge of the phenomenon. The research findings were published in a variety of journals, however, and for some time did not get the recognition that they deserved.

In the meantime, the Swedes had been working on the problem. In the 1940s, a soil scientist, Hans Egnér, decided to look at the way that crops were fertilised by nutrients from the atmosphere. He arranged for a number of sampling buckets to be located around Sweden, to measure atmospheric precipitation ('precipitation' here means rain, snow, sleet, hail, etc.) One of the factors to be checked was acidity.

Other scientists took up the idea, and expanded the network — first to Norway, Denmark, and Finland, and later to most of western and central Europe. This European Air Chemistry Network, as it came to be called, provided the first large-scale and long-term data on the nature of substances falling out of the air. In 1957, the network spread to include Poland and the USSR. With more than 100 collection stations, the system is still functioning now, and has been in continuous operation since its inception.[2]

Egnér was an agricultural scientist primarily concerned with the soil. Others were studying the atmosphere. Again in Sweden,

Erikkson and Rossby decided that certain substances could be carried both short and long distances through the air. This was not such an obvious hypothesis as it seems now. Although it had been realised since observations began that water is transported in clouds and rain, the idea that other materials could move around in a similar way was at the time unproven.

Using Egnér's analysis network, they tested various theories about the way the atmosphere picks things up and eventually returns them to the ground. On the basis of their observations, Erikkson formulated a general theory describing the way that matter circulates around the Earth. Supported by data from the European Air Chemistry Network, Rossby and Erikkson sponsored a series of conferences in the 1950s, in order to discuss atmospheric chemistry and dispersal processes. These conferences attracted scientists in many other fields of enquiry.

The first person to unify the different strands of acid rain research was Odén, a soil scientist at the Agricultural College near Uppsala in Sweden. In 1961, he started a monitoring network to check up on what was happening to Scandinavian lakes and streams ('surface water'). When information from this system was correlated with that from the European Air Chemistry Network and combined with Erikkson's work, a series of trends and relationships began to emerge.

Odén's analyses showed that:

- acid precipitation (acid-carrying rain, snow, hail, etc.) was occurring in many regions in Europe;
- both the precipitation and the lakes and streams in the suffering areas were becoming more acidic;
- a form of airborne import—export of sulphur- and nitrogen-carrying pollutants existed between countries, sometimes over very long distances—up to 2,000 kilometres (1,250 miles).

Odén also hypothesised that the probable consequences of acid input into soil and water over the years would be changes in the chemistry of lakes, decline in fish populations, release of poisonous metals from soils into surface water, decreased forest growth, increased plant diseases, and accelerated damage to materials—a formidable list of forecasts, most of which have subsequently proved to be correct.

Odén published his findings in two publications. In 1967, a prestigious Stockholm newspaper ran the story of his discoveries; and in 1968 a more scientific report appeared in an Ecology Committee *Bulletin*. The newspaper treated the findings as a kind of science-fiction horror scare, an insidious 'chemical war' between the industrial nations of Europe. It was a sensational story that evoked a wide public reaction. In much the same way, the Ecology Committee *Bulletin* stimulated interest throughout the scientific community.

The Swedish government eventually responded to these developments by pressurising the United Nations into discussing acid rain at the Conference on the Human Environment in 1972. Sweden presented a case study entitled, *Air Pollution Across National Boundaries: the Impact of Sulphur in Air and Precipitation.*

The size of the acidification problem was becoming more apparent, and a number of research projects were launched.

Norway

In Norway, the Norwegian Interdisciplinary Research Programme 'Acid precipitation—effects on forests and fish' began. The annual budget for this enquiry (known as the SNSF project) was $2 million per year from 1972 to 1980. It was a major scheme, designed to establish as precisely as possible the effects of acid precipitation on forests and freshwater fish, and to investigate the impact of air pollutants on soils, vegetation, and water. Eleven Norwegian institutions participated in the research programme, more than 150 scientists were engaged on the project over the years, and the final SNSF summary (in 1980) pulled together more than 300 reports and data analyses.

OECD

The second major initiative came from the Organisation for Economic Co-operation and Development (OECD), an association of the 24 leading industrial nations of the Western world.

The OECD programme on Long-Distance Transport of Air Pollutants (LRTAP) collected data from 11 European countries, in order to find out where the sulphur and sulphur dioxide was coming from, and where it was going to. The programme was published in 1977, and showed that the area of acid precipitation covers almost all of north-western Europe. It also confirmed that pollutants are often transported long distances, and demonstrated that the air quality in each European country is measurably affected by emissions from other countries in Europe.

After the OECD had obtained its results from LRTAP, it set out to discover what measures could be taken by member-states to prevent acid precipitation, and whether these were economically feasible. In 1981, the fruits of these deliberations appeared as *The Costs and Benefits of Sulphur Oxide Control,* the central methodology of which was to get around the numerous difficulties of air pollution analysis by working out three possible scenarios for sulphur dioxide emissions in 1985. Case I suggested that, in 1985, the European SO_2 emissions would be 121 per cent of the 1974 figure (this would happen if no desulphurisation programmes were started). Case II suggested that, if some desulphurisation was done, the 1985 figure would be 97 per cent of the 1974 level of emissions; and in Case III, in which a largish desulphurisation programme was undertaken, the 1985 figure would be 63 per cent of that in the control year. Then the OECD calculated the costs of limiting SO_2 levels in each of these projections, and weighed these costs against the environmental and financial benefits that would be gained. In both Case II and III, the report concluded, desulphurisation was economically feasible.

UNECE

The UN, meanwhile, had decided to get involved. Its Economic Commission for Europe (UNECE), in collaboration with two other bodies — the United Nations Environmental Programme and the World Meteorological Service — decided that it needed more information on origins and destinations of sulphur pollutants. It set up a scheme called the 'Co-operative Programme for Monitoring and Evaluation of Long-Range

Transmissions of Air Pollutants in Europe' (understandably shortened to 'European Monitoring and Evaluation Programme', or, even more conveniently, EMEP). The programme has some 50 stations in 18 countries in Europe, and they have been used to assess how much sulphur individual countries emit and receive. The first phase of the programme covered 1978 to 1980.

The UNECE also sponsored an international agreement to limit the export of air pollution. The 'Convention on Long-Range Transboundary Air Pollution' was signed by 33 countries in November 1979. This is a document deploring atmospheric pollution. Although it entails no binding obligations for reducing sulphur dioxide emissions, the signatories agree that they 'should endeavour to limit, and, as far as is possible, gradually reduce and prevent air pollution, including long-range transboundary air pollution'.

The convention entered into force in March 1983, and in the meantime, its executive body commissioned three extensive reports on the effects of sulphur compounds on (a) soil, ground water, and vegetation; (b) materials, including historic and cultural monuments; and (c) aquatic ecosystems (lakes, rivers, and streams), which were presented in September 1982.

Sweden

In the period 1972−82, the Swedes were working continuously on the acid problem, and in the early eighties, they decided to organise the 1982 Conference in Stockholm on the Acidification of the Environment, to mark the ten years' passage of time since their original evidence was presented to the UN. Preparatory to this, a committee of specialists wrote a background book, *Acidification Today and Tomorrow*, possibly the best book to date on the acid phenomenon. (The Swedes were so concerned to get their science right that they inserted 6,000 corrections before the final proof stage.) Shortly before the conference, an international panel of 103 scientific experts congregated in Stockholm to present working reports for use by the final assembly, which took place on 28−30 June 1982.

Canada and the USA

Large-scale research is being carried out in Canada and the USA. The Canadians are spending $25 million a year, employing 300 scientists, investigating the widespread pollution on their eastern coast and elsewhere; 350 projects were in progress by the end of 1983. In the USA, Congress passed the Acid Precipitation Act in 1980, providing for large-scale research within government agencies, within industry, and in international contexts. The USA was committed to spending $10 million a year for ten years on research into acidification, an amount which was subsequently increased.

The EEC

Finally, the EEC Commission and European Parliament have become involved. Although a Council Directive of 15 July 1980 had established limit values for air quality, and guide values for the amount of sulphur dioxide in the air, MEPs were still very concerned about environmental damage to their countries. As a response to this, the Environment Committee of the European Parliament had a two-day hearing on acid rain on 19 and 20 April 1983. A rapporteur, Herr Muntingh, was appointed to update knowledge of the extent of acid pollution. His researches provided substantial evidence of the scale of the threat. The EEC Commission also sponsored a three-day conference in Karlsruhe, on 19−21 September 1983, entitled 'Acid Deposition: A Challenge to Europe'.

This list of conferences and research studies appears here for a purpose − to establish that a large amount of work into the problem has already been carried out, in order to track down the mystery of where the pollution comes from, and what effects it has when it arrives. Discussion of the phenomenon in the UK seems to assume that acid rain is a new topic which has suddenly sprung up, and which needs to be subjected to thorough scientific examination before corrective measures can be undertaken.

The need for more scientific research is not denied − any environmental pollution that occurs on a scale as large as this should be a priority area for investigation − but on the other hand it should not be assumed, as we do in the UK, that little is known about the subject to date. A vast amount of knowledge has been accumulated.

The flow-chart shown in Figure A1 is an indication of the progression of the debate on acidification. It is a very simplified model, aiming to indicate some major European studies and it should not be taken as an exhaustive view. More than 3,000 studies into acidification have been carried out so far and the chart represents a small though significant fraction of the work to date.

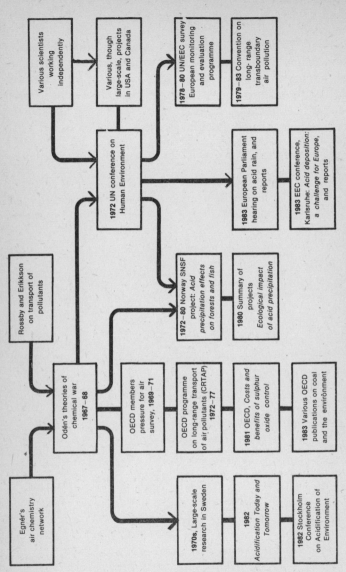

Figure A.1. More research needs to be done

Appendix B: Technologies for clean air

The technology for reducing sulphur and nitrogen oxides is well advanced; there are several systems in existence, some of which are already in commercial operation. The purpose of this appendix is to provide a glossary of terms rather than give a full explanation. For fuller information, the reader is referred to the Useful literature list on page 135.

Using low-sulphur fuel

Coal desulphurisation

The sulphur content of coal varies from 0.4−5.0 per cent by weight. European coal is at the lower end of the scale, ranging from France's 0.7 per cent to the UK's 1.5 per cent. (The figures here refer to hard coal; brown coal which is used in areas of Eastern Europe and the USA is not discussed.)

Hard coal contains both pyritic and organic sulphur, often in roughly equal proportions. At the moment, coal is crushed and cleaned in order to remove sand, clay, and other incombustible particles. This process also extracts a small amount of pyritic sulphur.

The system can be optimised to remove up to 90 per cent of the pyrite, although average removal is nearer 50 per cent − that is, an extraction of 25 per cent of the total sulphur content of the coal. The OECD has described a process based on differences in specific gravity, which can remove 30−35 per cent of the total sulphur content. This would cost between £1 and £2.50 per ton of coal. (The organic sulphur, mentioned above, is difficult to extract, so strategies at the moment concentrate on the pyrite.)

If this extraction figure and cost is applied to the UK electricity industry's output, then the SO_2 emitted from our coal-powered power stations would have been in the order of 1,526,000 tonnes instead of 2,300,000 tonnes for 1982−83. This is a very rough calculation, but it represents an SO_2 reduction of 33 per cent at a cost of 2.5−4.6 per cent in the price of the power station coal. The reduction for the UK as a whole would be 18 per cent.

These figures are not, it must be said, confirmed by research currently being done by the NCB. Their Mining Research and Development Establishment, in response to a request from the CEGB, carried out feasibility tests on coal desulphurisation in 1983. Coal supplied to power stations has a higher sulphur content than the national average of 1.5 per cent−the CEGB average is 1.65 per cent, of which 0.8 per cent is organic, 0.85 per cent pyritic. The Mining Research and Development Establishment identified the 24 collieries supplying the highest sulphur-content coal (about 1.9 per cent) and tried to work out the practicability of removing sulphur before the coal is used. At the time of going to press, they would not release the results of their research, saying only that the systems needed are 'different for every coal, and each coal has to be examined individually'. Indications of the results of their investigations were that 10−15 per cent of sulphur could be removed from the coal, with a possible maximum of 20 per cent.

Oil desulphurisation

Crude petroleum can be refined to produce light fuel oils with 0.2−0.7 per cent sulphur by weight, and heavy oils with 3−5.5 per cent sulphur. Combustion of these fuels results in sulphur dioxide emissions, with most (about 80 per cent) coming from the heavy oils.

The sulphur content of these heavy oils can be reduced by widely used processes in modern refineries. In a study presented to the 1982 Stockholm Conference on the Acidification of the Environment, a process for desulphurising was described which was capable of decreasing the sulphur content of heavy fuel oils to 0.5 per cent.

The report, presented by the Air Pollution Control Directorate of Environment Canada, concluded that this strategy is cheaper than cleaning the smoke coming out of the chimneys — though to minimise costs, the operation should be carried out on as large a scale as possible.

This kind of approach — desulphurising oil before it is sold — has been used in Japan (see page 67).

Cleaning the smoke as it comes out of the chimney

This is commonly referred to as Flue-Gas Desulphurisation, or FGD, and is the most immediate answer to a sulphur-polluting plant. There are various kinds of systems, and a fair amount of jargon — wet and dry scrubbers, throwaway, regenerative, or gypsum processes — but the technology involved is important and bears closer examination.

'Wet' and 'dry' refer to the two types of FGD, the former using large quantities of water in the cleaning process; and 'throwaway', 'regenerative' and 'gypsum' refer to the end-product that is left when the cleaning is finished. A throwaway system results in large quantities of sludge that have to be dumped somewhere. Regenerative FGD produces sulphur or sulphuric acid, which can then be marketed. The third process produces gypsum (dihydrated calcium sulphate), a mineral substance from which, in its natural occurrence, plaster of Paris and other plasters can be obtained. It can also be used in the construction industry.

FGD technology depends on the reaction of flue gases with a chemical absorbent, a process known as 'scrubbing'. In wet scrubbing, the absorbent is mixed with water and then sprayed on to the gases. A number of different substances can be used, but the most common are lime or limestone.

Wet FGD 1: Throwaway
Lime or limestone mixed with large quantities of water is sprayed on to the smoke coming out of the combustion area. Sulphur dioxide in the gases combines with the lime/limestone slurry to produce calcium sulphite, in the form of a sludge.

The quantities of liquid waste produced by this system can

cause difficulties. It is cumbersome to handle and transport, and also presents problems of disposal. The annual sludge production from a 2,000 MW power station could exceed 300,000 m³ (66 million gallons). If there are no landfill sites nearby, present practice is to pond the wastes in pits—not a very satisfactory solution to a pollution problem.

This is the prevailing technique in the USA, and limestone rather than lime is used because of its lower cost.

Wet FGD 2: Gypsum

Because of the waste-disposal problem presented by the slurry sludge, most new FGD systems entering operation or in the planning stage, especially in Europe and Japan, use air oxidation to change the calcium sulphite into calcium sulphate-dihydrate (gypsum). This still poses waste-disposal difficulties unless the product is used commercially (current estimates for gypsum production in these systems amount to 1.5 to 2 million tonnes per year from 1990).

In Japan, where there is little natural gypsum, the FGD product is purified and marketed. In West Germany, the prognosis also looks good. In 1978, the gypsum used in Germany (95 per cent of which was taken from natural sources) amounted to 5 million tonnes—3 million used as construction material, the rest added to cement as a hardener. FGD gypsum could thus meet a large market. Current practice in West Germany is generally for manufacturers of FGD to offer sales guarantees for by-product gypsum when they sell their systems.

The throwaway and gypsum processes are at the moment superior to other systems, both technically and from the economic point of view. SO_2 removal up to 95 per cent can be achieved. For these reasons they are the most frequently used.

Wet FGD 3: Regenerative

These are designed to produce sulphuric acid or sulphur. According to a 1983 OECD report, two commercial techniques exist but economic considerations have limited their application. A new process has just been announced (the Mark 13A) which produces sulphuric acid and hydrogen. The disposal of solid wastes and water is not required. It is too early to say whether this system has commercial applications.

Dry FGD

This uses much less water than wet systems. An alkaline absorbent slurry is sprayed on to a hot SO_2-carrying flue-gas stream. As the slurry is dried by the hot gas, SO_2 is trapped, producing a dry powder mixture which is collected in filter bags.

This technology is being adopted by some power plants, principally in the USA, for use with low-sulphur fuel where small amounts of sulphur dioxide need to be extracted.

Conclusion

'Throwaway' FGD is currently the most common system, though 'gypsum' is rapidly being adopted as an environmentally sounder option. Both 'regenerative' and 'dry' FGD are regarded as having possible applications in the future, and some dry FGD is in use at the moment.

Flue-Gas Desulphurisation is suitable for large emission sources, but the capital and operational costs make it impractical for small fuel-users. A national desulphurisation strategy would entail a combination of this system with preparation of low-sulphur coal and oil. High-sulphur fuel could be used by FGD sources such as power stations, in which 95 per cent of the sulphur could be removed. The low-sulphur coal and oil could then be routed to small industries and users which could not afford FGD technology.

Redesigning the furnaces in which the fuel is burnt

The advantage of this approach to the question of desulphurisation is that, not only does it decrease the emissions of SO_2, but also of NO_x, by lowering the combustion temperature (see the section on NO_x, page 130). A long-term response to the need for cutting SO_2 and NO_x emissions will have to consider changing the furnaces in which fossil fuels are burnt. The new systems should not be regarded as the solution to the current acidification crisis, however. It would take at least ten years to plan, build, and get new furnaces into operation, by which time it is difficult to forecast the consequences of continued acidic pollution. As described above, immediate action to combat acid rain involves desulphurising fuel used in industry and power stations, or cleaning the smoke from the chimneys. The systems described here should be regarded as plans for the future. Information is taken from *Acidification Today and Tomorrow* and *Strategies and Methods to Control Emissions of Sulphur and Nitrogen Oxides* (see Useful literature, page 135).

Lime Injection in Multi-Stage Burners (LIMB)
Where powdered coal is used as the fuel, powdered limestone can be injected into the firebox and the combustion temperature lowered by the use of special burners, flue-gas reflux or multi-stage combustion. Uncertain findings from isolated tests indicate that 50−70 per cent of the sulphur can be fixed into the ash as gypsum.

Capital costs of the technology are low, which makes it interesting for plants with short operational lives. As yet, however, this has not been employed on a commerial scale.

Combustion modification by the use of multi-stage burners is recognised as 'state of the art' to reduce up to 50 per cent of NO_x emissions compared with emissions from conventional burners. In the case of coal, the additional cost of combustion modifications is generally less than 1 per cent of the total power plant capital cost.

Development of low NO_x burners with limestone injection systems aiming at simultaneous reduction of sulphur and nitrogen oxides is in the early research stage and has shown promising results. Such technology has interesting potential because of its simplicity and flexibility for use in industrial installations, and also as retrofit (that is, fitting new technology to existing plant, as opposed to having it included from the initial stages of planning). Pilot test results on hard coal have indicated removal efficiencies up to 80 per cent for SO_2 and 50 per cent for NO_x.

Atmospheric, bubbling fluidised bed
In this system, the fuel (coal) and various non-combustible

materials are put into a vessel with a perforated bottom. Most of the bed material is coal ash together with limestone, sand, or similar materials that can absorb sulphur. The fuel itself comprises only 1−5 per cent of the entire bed.

Air is blown into the vessel and through the bed, causing the particles to float freely. The bed behaves as a fluid−hence the term 'fluidised bed'. Most of these systems function like ordinary boilers, with atmospheric pressure in the combustion area. If the pressure is raised by air compression to 5−20 times atmospheric pressure, several advantages may occur.

Combustion takes place at about 850°C and without visible flames. The fuel is fed continuously into the bed together with the non-combustible material (e.g. limestone); the final products−gypsum and limestone−are discharged at the same rate.

Advantages of fluidised beds include: reduction of SO_2 emissions as the sulphur is fixed to the limestone; low combustion temperatures which keep down NO_x emissions; a technology which is relatively immune to variations in fuel quality; and boiler units which are compact and thus cheaper.

Generally 80−90 per cent of the sulphur can be removed, although this is dependent on the temperature−SO_2 emissions may be doubled if the temperature is raised by 15°C.

Atmospheric, high-speed fluidised bed

This uses the same principle as an ordinary fluidised bed, but the bed material circulates very quickly. It is the most highly-developed technology and is available for commerical use in sizes up to 250 MW thermal rating, though larger sizes are being developed. The operational reliability of one plant, at Kauttua in Finland, has proved to be satisfactory.

Fluidised-bed plants call for the same capital investment as for conventional boilers. The costs of sulphur removal are, by and large, the cost of the limestone, its transport and waste handling. The cost of the limestone is about a 2 per cent surcharge on the cost of fuel. Oil can be burned in this system, but so far only coal has been used.

Bubbling fluidised beds have been in use for many years in

smaller installations. The circulating fluidised bed system is now in commercial operation in several plants in Finland. Pressurised fluidised beds are being tested in Sweden, West Germany, and the USA.

Cutting down on NO_x

Although — compared with the amount of SO_2 produced — NO_x emissions at 9 million tonnes a year (OECD Europe) are relatively small, they are increasing rapidly. According to the United Nations Economic Commission for Europe, 'emissions of NO_x are expected to undergo major increases over the next two decades in both Europe and North America'. Norwegian research estimates that nitric acid contributes about 30 per cent of acid precipitation, and the proportion is increasing. NO_x is also one of the originators of ozone, which at ground level has caused crop damage in the USA and is, according to some observers, responsible for large-scale damage to trees in Germany. The synergistic effects of SO_2, NO_x and ozone (see page 47) are still being examined. SO_2 is the substantial contributor to acidification, but NO_x is increasingly being recognised as having great importance in the process. NO_x emissions should accordingly be reduced.

Scientific consensus holds that 50 per cent of NO_x is produced from motor traffic and 50 per cent from industry, but the figures vary from country to country. In the UK, approximately 28 per cent of NO_x is produced by traffic, with the rest coming from stationary sources (power stations and industry); power plants produce more NO_x than any other source — about 45 per cent in 1981. The picture is similar in West Germany, but different in Australia, Japan, and the USA, where No_x is jointly created by vehicles and stationary sources, and in Canada and Scandinavia, where traffic is the main contributor.

Reducing NO_x in industrial smoke involves lowering the temperature at which the coal or oil is burnt. Nitrogen oxide emissions result from two sources of nitrogen introduced into the combustion process. Firstly, nitrogen in the air is carried along with oxygen into the combustion chamber. At high temperatures, a proportion of this reacts with the oxygen to form nitrogen oxides.

Secondly, there is some nitrogen in the fuel. When oil is burnt, about half of the nitrogen oxides that result come from the fuel itself. The corresponding figure for coal is about 80 per cent.

Strategies for reducing NO_x concentrate on adjusting the way in which combustion takes place, to lower the temperature and pressure of the burning process, as well as slowing down the mixing rate of the fuel and air. This reduces the intensity of combustion and thus the amount of NO_x formed.

There are a number of denitrification plants in operation, almost wholly in Japan, where the most highly developed system is selective catalytic reduction (SCR), which uses ammonia to reduce NO_2 to nitrogen in a catalytic reaction. Large-scale flue-gas denitrification systems have been working in Japan since 1974, with NO_x removal efficiencies of up to 90 per cent.

Five countries (Australia, Canada, West Germany, Japan and the USA) have specified NO_x emission limits from stationary sources. In Japan, regulations apply to some 13,000 facilities producing nitrogen oxides; as a result, total NO_x emissions in the country have been reduced by about 30 per cent since 1978 when there were no restrictions. According to the UK Department of Environment, there are no plans for similar legislation in the UK at the moment.

Traffic

The amount of NO_x pollution from traffic is increasing. The most efficient way to reduce vehicular NO_x emissions is to fit a three-way catalytic converter to the car exhaust. This drastically cuts emissions of carbon monoxide, hydrocarbons, and nitrogen oxides. It also uses lead-free petrol, which is why West Germany is going to abolish lead in its fuel from January 1986, regardless of protests from German car manufacturers and from other EEC countries who want to move towards lead-free petrol more slowly.

The catalytic converter has produced good results in Canada and Japan, and catalysts have been employed in the USA and Japan for some years. Research on the system is still not complete, but the costs of fitting have been estimated at between £150 and £300 per car, using existing technology. Fuel consumption penalties of up to 10 per cent may be incurred, though this is by no means definite.

Appendix C: Further information

Campaigning groups

United Kingdom

Acid Rain Clearing House (ARCH)*
Friends of the Earth Scotland
53 George IV Bridge
Edinburgh 1
tel: 031 - 225 6906

Friends of the Earth Ltd*
377 City Road
London EC1V 1NA
tel: 01 - 837 0731

Greenpeace*
36 Graham Street
London N1
tel: 01 - 251 3020

Ecology Party*
36/38 Clapham Road
London SW9 OJQ
tel: 01 - 735 2485

Socialist Environment and Resources Association*
9 Poland Street
London W1
tel: 01 - 439 3749

Young Liberals Ecology Group*
1 Whitehall Place
London SW1
tel: 01 - 839 2727

National Society for Clean Air
134 North Street
Brighton 1
tel: 0273 - 26313

Scottish Wildlife Trust
25 Johnstone Terrace
Edinburgh EH1
tel: 031 - 226 4602

World Wildlife Fund
Panda House
Ockford Road
Godalming
Surrey

*These six groups have been co-operating in a Stop Acid Rain UK campaign.

British Association of Nature Conservationists
Rectory Farm
Stanton St John
Oxford OX9 1HF
tel: 086 735 - 214

Earth Resources Research
258 Pentonville Road
London N1
tel: 01 - 258 3833

International

Stop Acid Rain Sweden
The Swedish NGO Secretariat on Acid Rain
The Swedish Society for the Conservation of Nature
Box 6400
S-113 82
Stockholm
Sweden
tel: Stockholm 15 15 50

Stop Acid Rain Norway
PO Box 8268
Hammersborg
N-Oslo 1
Norway
tel: Oslo 42 95 00

Die Grunen (The Greens)
Colmanstrasse 36
D/5300
Bonn 1
West Germany
tel: Bonn 692021 or 692022

Canadian Coalition on Acid Rain
112 St Clair Avenue West
Suite 504
Toronto
Ontario M4V 2Y3
tel: Toronto 968 2135

The Acid Rain Foundation Inc.
1630 Blackhawk Hills
St Pauls
Minnesota 55122
USA

Useful contacts

International bodies who have been researching acid rain

United Nations Economic Commission for Europe
Palais des Nations
CH-1211 Geneva 10
Switzerland

United Nations Environment Programme
Palais des Nations
CH-1211 Geneva 10
Switzerland

UNESCO
7 place de Fontenoy
75700 Paris
France

OECD
2 rue Andre-Pascal
75775 Paris Cedex 16
France

EEC Commission
Research, Science and
 Development
Rue de la Loi 200
B-1049 Brussels
Belgium

Useful literature

General

Environment '82 Committee, *Acidification Today and Tomorrow*,
 published by Swedish Ministry of Agriculture, 1982. (Highly
 recommended.)
Environment Canada, *Downwind: The Acid Rain Story*, Minister of
 Supply and Services, Canada 1982.
Environmental Resources Ltd, *Acid Rain: A Review of the
 Phenomenon in the EEC and Europe*, Graham and Trotman
 1983.
Herr Muntingh, *Information Note on Acid Rain*, European
 Parliament, 1983.
Acid Rain Foundation Inc., *Acid Rain Resources Directory*, April
 1983.

Articles

'Rain more acid than vinegar', *New Scientist*, 12 August 1982
 (special acid rain issue).
R. Hoyle, 'The silent scourge', *Time*, 8 November 1982.
A. Tucker, 'Britain's two million tonne chemical warfare
 onslaught', *Guardian*, 22 September 1983.
G. Lean, 'British lakes poisoned by acid rain', and 'Acid rain kills
 fish in Welsh rivers', *The Observer*, 18 and 25 September 1983.
B. Moynahan, 'The rain that kills', *Sunday Times*, 5 June 1983.
Acid News, available every two months from Stop Acid Rain
 Sweden (see 'Useful contacts').

More technical

OECD, *The Costs and Benefits of Sulphur Oxide Control*, 1981.
- *Coal: Environmental Issues and Remedies*, 1983.
- *Coal and Environmental Protection*, 1983.
- *Control technology for Nitrogen Oxide Emissions from Stationary Sources*, 1983.
- *Costs of Coal Pollution Abatement*, 1983.
- *Programme on Long-Range Transport of Air Pollutants*, 1979.

National Swedish Environment Protection Board, *Ecological Effects of Acid Deposition*, and *Strategies and Methods to Control Emissions of Sulphur and Nitrogen Oxides*, 1982.

Commission of the European Communities, *Acid Deposition: A Challenge for Europe*, 1983.

National Research Council USA, *Acid Deposition: Atmospheric Processes in Eastern North America*, National Academy Press, 1983.

SNSF project, 1972−80, *Acid Precipitation − Effects on Forests and Fish*, 1981, and *Ecological Impact of Acid Precipitation*, 1980.

UK Review Group on Acid Rain, Warren Spring Laboratory, *Acid Deposition in the United Kingdom*, 1983.

Notes

1. Acid words

1. UK Review Group on Acid Rain, *Acid Deposition in the United Kingdom*, Stevenage: Warren Spring Laboratory 1984, p.1.
2. Environment Canada, *Downwind: The Acid Rain Story*, Ottawa: Ministry of Supply and Services, Canada 1982, p.9.
3. United Kingdom Review Group on Acid Rain, *op.cit.* p.55.
4. Environmental Resources Ltd, *Acid Rain*, London: Graham and Trotman 1983, p.5. This figure varies according to distance from pollutant source, local industries, and so on and should be regarded as an indication of relative proportions rather than a final figure.
5. L. Overrein, H.M. Seip and A. Tolla, *Acid Precipitation. Effects On Forest and Fish*, Oslo: SNSF project final report 1980, p.17.
6. *Ibid.* p.17. Also Committee on Atmospheric Transport and Chemical Transformation in Acid Precipitation, National Research Council, *Acid Deposition: Atmospheric Process in Eastern North America*, Washington: National Academy Press 1983, p.13.
7. D.W. Sutcliffe and T.R. Carrick, *Acid Rain and the River Duddon in Cumbria*, Ambleside: The Freshwater Biological Association 1983, p.2.
8. Overrein, Seip and Tolla, *op.cit.* p.18.
9. *Ibid.* p.19. Also, B. Ottar, 'Air pollution, emission, and ambient concentrations', *Acid Deposition: A Challenge for Europe*, Karlsruhe: EEC 1983, p.36.
10. Overrein, Seip and Tolla, *op.cit.* p.19, and Environmental Resources Ltd, *op.cit.* p.7.
11. Environmental Resources Ltd, *op.cit.* p.53.

12. M.K. Tolba, *The State of the World Environment Report 1983*, Geneva: United Nations Environment Programme 1983, p.7.
13. Information in Figure 1.2 is from Ottar, *op.cit.* p.39, and from Environment '82 Committee, *Acidification Today and Tomorrow*, Stockholm: Swedish Ministry of Agriculture 1982, pp.32 and 33.
14. United Nations Economic Commission for Europe, *Report on Effects of Sulphur Compounds on Materials*, Geneva: 1982, p.24. Corrosion damage is estimated at $4.50 per person per year, with the population of Europe taken as 680 million. The report notes that this figure is probably an underestimate.
15. Where sterling equivalents are given, this is done by adjusting for inflation in the currency concerned, then translating them at 1984 rates of exchange. This is not intended as a precise translation, but merely a rough equivalent; exchange value at $1.50 to £1 sterling.
16. Tolba, *op.cit.* p.1.
17. *Guardian*, 20 January 1984. The figure is taken from European Parliament Working Document 1-1168/83: damages estimated at 3 – 5 per cent of European Community Gross National Product. The report notes that this is likely to be an underestimate.

2. Acid effects

1. F.H. Bormann, 'Factors confounding evaluation of air pollution stress on forests: pollution input and ecosystem complexity' in *Acid Deposition: A Challenge for Europe*, Karlsruhe: EEC 1983, p.150.
2. S. Mills, 'Forestry in Britain: planting for alternative futures', *New Scientist*, 4 August 1983, p.336.
3. F.T. Last, 'Effects of atmospheric sulphur compounds on natural and man-made terrestrial and aquatic ecosystems', *Agriculture and Environment*, no.7, 1982, p.343.
4. B. Ulrich, Effects of accumulation of air pollutants in forest ecosystems', in *Acid Deposition: A Challenge for Europe*, Karlsruhe: EEC 1983, p.132.
5. Bormann, *op.cit.* pp.151 – 2.
6. European Communities R & D Information No. 23, *Acid Rain – A Challenge for Europe*, Brussels: 1983, p.5.
7. *New Scientist*, 26 October 1983.
8. B. Moynahan, 'The rain that kills', *Sunday Times*, 5 June 1983, p.13.

9. European Communities R & D Information, *op.cit.* p.5.
10. *Ibid.* p.5. Converted from ECUs in the original: 1 ECU taken as 57p.
11. Statement made by Professor Ulrich at 1983 EEC Karlsruhe conference on acid rain.
12. European Parliament Working Document 1-1168/83, *Report on the Combating of Acid Rain*, 1983, p.27.
13. United Nations Economic Commission for Europe (UNECE), *Report on Effects of Sulphur Compounds on Soil, Ground Water and Vegetation*, Geneva:1982, p.10.
14. Committee on the Atmosphere and the Biosphere, *Atmosphere — Biosphere Interactions: Towards a Better Understanding of the Ecological Consequences of Fossil Fuel Combustion*, Washington: National Academy Press, 1981, p.2.
15. Environment '82 Committee, *Acidification Today and Tomorrow*, Stockholm: Swedish Ministry of Agriculture 1982, p.55.
16. Committee on the Atmosphere and Biosphere, *op.cit.* p.3.
17. Environment '82 Committee, *op.cit.* p.67.
18. H. Hultberg, 'Effects of acid depositions on aquatic ecosystems', *Acid Deposition: A Challenge for Europe*, Karlsruhe: EEC 1983, p.174.
19. P.G. McWilliams, 'A comparison of physiological characteristics in normal and acid exposed populations of the brown trout *Salmo trutta*', *Comp. Biochem. Physiol.*, vol. 72a, no. 3, 1982, p.515.
20. Hultberg, *op.cit.* p.171.
21. W. Dickson, 'Water acidification — effects and countermeasures', *Ecological Effects of Acid Deposition*, Stockholm: National Swedish Environment Protection Board 1982, p.272.
22. I.P. Muniz, 'The effects of acidification on Norwegian freshwater systems', *Ecological Effects of Acid Deposition*, Stockholm: National Swedish Environment Protection Board 1982, p.315.
23. Hultberg, *op. cit.* p.177.
24. A. LaBastille, 'Acid rain: how great a menace?', *National Geographic*, vol. 160, no. 5, November 1981, p.673.
25. From Environment '82 Committee, *op. cit.* p.44.
26. UK Review Group on Acid Rain, *Acidity of Rainfall in the United Kingdom — a Preliminary Report*, Stevenage: Warren Spring Laboratory 1982, p. 27.
27. Muniz, *op. cit.* p.317.

28. L. Overrein, H.M. Seip and A. Tollan, *Acid Precipitation. Effects on Forest and Fish*, Oslo: SNSF project final report 1980, p.161.
29. European Communities R & D Information, *op. cit.* p.5.
30. Environmental Resources Ltd, *Acid Rain*, London: Graham and Trotman 1983, p.129. This appears to be an underestimate based on 1974 OECD calculations in *The Costs and Benefits of Sulphur Oxide Control*, 1981, p.105.
31. Environmental Resources Ltd, *op. cit.* p.129.
32. I.P. Muniz, *op. cit.* p.317.
33. The speech was made I believe to the Canadian Chamber of Commerce. Information supplied by Canadian High Commission, London.
34. Information given by Dr Morrison of the Department for Agriculture and Fisheries for Scotland in a conference on acidification at University College London, 16 December 1983. (See p.91.)
35. Conference organised by Socialist Environment and Resources Association: comment transcribed from tape-recording.
36. Information from *Urban Change and Conflict — Pollution in Cracow*, (TV 6) Open University, pp.96-8, and R. Serafin, 'The greening of Poland', *The Ecologist*, vol. 12, no. 4, 1982.
37. *New Scientist*, 21 April 1983, p.196.
38. United Nations Economic Commission for Europe (UNECE), *Report on the Effects of Sulphur Compounds on Materials, including Historic and Cultural Monuments*, Geneva: 1982, p.11.
39. *Ibid.* pp.11, 15, 16.
40. *Ibid.* p.20.
41. Herr Muntingh, *Information Note on Acid Rain*, Brussels: European Parliament Committee on the Environment, Public Health and Consumer Protection 1983, p.6.
42. *Ibid.* p.5.
43. *Ibid.* p.5.
44. J. May, 'Dragon's revenge', *Undercurrents*, no. 54, 1982, p.9.
45. Muntingh, *op. cit.* p.6.
46. United Nations Economic Commission for Europe, *Effects of Acid and Acidifying Compounds on Materials used in Buildings, Monuments and Artifacts of Historical and Cultural Significance*, Geneva: January 1984, p.2.
47. May, *op. cit.* p.9.

48. *Acid News*, October 1983, p.13. Corrosion and repair costs confirmed by London New York Information Service, also referred to in LaBastille, *op. cit.* p.674.
49. R. Hoyle, 'The silent scourge', *Time*, 8 November 1982, p.42.
50. May, *op. cit.* p.9.
51. T.N. Skoulikidis, 'Effects of primary and secondary pollutants and acid depositions on (ancient and modern) buildings and monuments', *Acid Deposition: A Challenge for Europe*, Karlsruhe: EEC 1983, p.210.
52. European Parliament, *Summary of the Public Hearing on Acid Deposition, 19 and 20 April 1983*, Brussels: p.28.
53. Skoulikidis, *op. cit.* p.208.
54. UNECE, *op. cit.* (note 38) p.25.
55. OECD, *The Costs and Benefits of Sulphur Oxide Control*, Paris: 1981, p.16.
56. *Ibid.* p.76.
57. *Ibid.* pp.69 and 75.
58. *Ibid.* p.16.
59. UNECE, *op. cit.* (note 38) p.29.
60. Statement made by Herr Muntingh at press conference after hearing. Quotation taken from notes.
61. World Health Organisation (WHO), *Sulphur Oxides and Suspended Particulate Matter*, Geneva: WHO, p.4.
62. OECD, *op. cit.* p.90.
63. WHO, *op. cit.* p.5.
64. OECD, *op. cit.* p.92.
65. *Ibid.* pp.98-9.
66. WHO, *op. cit.* pp.6-7.
67. M.J.R. Schwar, *Smoke and Sulphur Dioxide Levels in Greater London, Period Summer '72 to Winter '77*, London: GLC Scientific Branch 1979, p.1.
68. EEC *Council Directive on Air Quality Limit Values for Sulphur Dioxide and Suspended Particulates, 15 July 1980*, Articles 2 and 5.
69. GLC, *A Review of Air Pollution by Sulphur Dioxide and Smoke in London*, 1982, p.5.
70. *Ibid.* p.8.
71. E.W. Culley, R. Evans and C. Armogie, *Summary of Sulphur Dioxide and Smoke Concentrations for Years 1981/2 and 1982/3*, London: GLC 1983.
72. Telephone conversation.

73. GLC, *op. cit.* p.5.
74. M. Furuichi, 'Sulphur reduction policy in Japan', *Technocrat*, vol. 11, no. 9, September 1978, p.29.
75. *Observer*, 20 November 1983.
76. Kawecki Associates, *Sulphur Oxides and Public Health: Evidence of Greater Risks*, New York: American Lung Association 1983.
77. *New Scientist*, 9 February 1984.
78. *International Herald Tribune*, 4 January 1984.
79. WHO, *op. cit.* p.6.
80. C.R. Creekmore, 'If you breathe, don't smoke', *Science 84*, January/February 1984, p.84.
81. *Der Spiegel*, 9 January 1984.
82. R. Hoyle, 'The silent scourge', *Time*, November 1982, p.44.
83. F.T. Last, 'Direct effects of air pollutants, singly and in mixtures, on plants and plant assemblages', *Acid Rain: A Challenge for Europe*, Karlsruhe: EEC 1983, p.114.
84. F.T. Last, 'Effects of atmospheric sulphur compounds on natural and man-made terrestrial and aquatic ecosystems', *Agriculture and Environment*, vol. 7, 1982, p.317.
85. Environment '82 Committee, *op. cit.* p.84.
86. OECD, *op. cit.* p.116.
87. *Sunday Standard*, 6 March 1983.
88. *Farmers' Weekly*, 13 January 1984.
89. *Press Advisory Release: Interim Report from Acid Rain Peer Review Panel*, Washington: Executive Office of the President, 28 June 1983.
90. Environmental Resources Ltd, *op. cit.* p.85.
91. *New Scientist*, 9 February 1984.
92. *Taraxacum*, no. 1, 1982, p.2.

3. Acid politics

1. J.W. McNeill, 'Coal and environment: constraint or an opportunity', *Costs of Coal Pollution Abatement*, Paris: OECD 1983, p.57.
2. Figures taken from the Highton and Chadwick chart, Table 3.1.
3. Environment '82 Commitee, *Acidification Today and Tomorrow*, Stockholm: Swedish Ministry of Agriculture 1982, p.41. Information confirmed verbally from UK Meteorological Office, although the complications of European wind patterns were

pointed out in a letter that followed. Over the UK the prevailing wind is south-west, south, or west; over Denmark, southern and central Sweden and Finland, winds are westerly and south-westerly all year; in France, Belgium, the Netherlands and Luxembourg, during winter (period of maximum pollution) most frequent wind directions are between south and west in the north of the area, and in parts of southern France between west and north; in mediterranean regions winds are north to north-west.

4. N.H. Highton and M.J. Chadwick, 'The effects of changing energy patterns of energy use on sulphur emissions and depositions in Europe', *Ambio*, vol. XI, no. 6, 1982.
5. Telephone conversation with the Department of Environment.
6. *New York Times*, 8 June 1980.
7. Commission on Energy and the Environment, *Coal and the Environment*, London: HMSO 1981, p.153.
8. See page 95.
9. See page 87.
10. United Nations Economic Commission for Europe, *Convention on Long-Range Transboundary Air Pollution*, 13 November 1979.
11. *Report and Background Papers, 1982 Stockholm Conference on the Acidification of the Environment*, Stockholm: National Swedish Environment Protection Board 1982, p.20.
12. Statement by EEC Council of Ministers, March 1983: 'The damage done to the forestry environment by acid rain makes effective joint action urgently necessary.'
13. *New Scientist*, 15 September 1983, p.747.
14. *Guardian*, 22 September 1983, p.19.
15. *Observer*, 18 and 25 September 1983.
16. *New Scientist*, 15 September 1983.
17. EEC, *Proposal for a Council Directive on Limiting Emissions from Large Combustion Plants*, 15 December 1983.
18. French Ministry of Environment Press Release, 22 February 1984.
19. Giles Shaw, Parliamentary Under-secretary of the Environment, in the House of Commons, July 1982.
20. Patrick Jenkin, Secretary of State for the Environment, in the House of Commons, 8 February 1984.
21. European Communities R & D information No. 23, *Acid Rain − A Challenge for Europe*, Brussels: 1983, p.3.
22. M.K. Tolba, *The State of the World Environment Report 1983*,

Geneva: UNEP 1983, p.7.

23. Commission on Energy and the Environment, *op. cit.* p.154. The 1983 figure comes from *Digest of Environmental Pollution and Water Statistics*, no. 6, London: HMSO 1983.

24. *Royal Commission on Environmental Pollution, Tenth Report*, London: HMSO 1984, p.145.

25. B. Ottar, 'Air Pollution, Emission and Ambient Concentrations', *Acid Deposition: A Challenge for Europe*, Karlsruhe: EEC 1983, p.36.

26. *Digest of Environmental Pollution and Water Statistics*, no. 5, London: HMSO 1982, p.14.

27. Ottar, *op. cit.* p.37.

28. B. Scharer and N. Haug, 'On the economics of flue gas desulphurisation', *Costs of Coal Pollution Abatement*, Paris: OECD 1983, p.135.

29. OECD, *Coal: Environmental Issues and Remedies*, Paris: 1983, p.64.

30. Verbal communication from CEGB official.

31. *Guardian*, 2 March 1984. Representations made by Denmark, Finland, Iceland, Greenland, Sweden and Norway.

32. *Royal Commission on Environmental Pollution, Tenth Report*, London: HMSO 1984, p.145.

33. N. Highton, *The Impact of SO_2 Control on Electricity Prices*, London: paper presented at the Watt Committee on Energy, 1 December 1983, p.6.

34. *Ibid.* p.6.

35. Friends of the Earth, *Acid Rain: A Friends of the Earth Special Report*, London: 1984, p.4. It may be worth comparing this with an OECD estimate three years earlier of electricity price increases of 2.5−3.5 per cent (OECD, *The Costs and Benefits of Sulphur Oxide Control*, Paris: 1981, p.15).

36. Letter from CEGB Secretary to *The Times*, 2 April 1983.

37. Testimony from Sizewell Inquiry, *Guardian*, 12 November 1983.

38. Testimony from Sizewell Inquiry, *Guardian*, 22 October 1983.

39. OECD, *The Costs and Benefits of Sulphur Oxide Control*, Paris: 1981, p.20.

40. OECD, *Coal and Environmental Protection*, Paris: 1983, p.115: 'It is clear that the current ratio of coal and oil prices is such as to make oil-plant uncompetitive.' Also CEGB, *Annual Report and Accounts 1982−83*, p.11.

41. *Britain's Energy Resources*, London: HMSO 1982, p.17.
42. E.S. Rubin, *'Summary and analysis', Costs of Coal Pollution Abatement*, Paris: OECD 1983, p.24: 'Across OECD countries the cost of a nuclear plant may vary by a factor of two (from $1,200 to $2,000 per kW) depending on applicable safety standards and other cost factors, thus making nuclear power either the cheapest or the most expensive option relative to fossil fuels.' See also OECD, *Coal and Environmental Protection, op. cit.* p.114: 'In the past decade, nuclear capital costs have risen in real terms by 10—15 per cent per annum.'
43. *Guardian*, 4 February 1984.
44. H. Thomas, 'A chill of complacency over Britain's energy saving', *Guardian*, 7 February 1984.
45. *Guardian*, 11 January, 1984.
46. Testimony from the Sizewell Inquiry, *Guardian*, 12 November 1983.
47. OECD, *Coal: Environmental Issues and Remedies, op. cit.* p.86.
48. Quoted in *The Times*, 19 November 1953.
49. *New Scientist*, 5 January 1984.
50. H. Hamanaka, 'Environmental measures for the expanded use of coal in Japan', *Costs of Coal Pollution Abatement*, Paris: OECD 1983, p.85.
51. *Ibid.* p.87.
52. *Ibid.* p.67, and M. Furuichi, 'Sulphur Reduction Policy in Japan', *Technocrat*, vol. 11, no. 9, September 1978, pp.20—2.
53. Furuichi, *op. cit.* p.26.
54. *Ibid.* p.32.
55. Hamanaka, *op. cit.* p.91.
56. T. Kojima, 'Revision of the environmental quality standard for nitrogen dioxide', *Technocrat*, vol. 11, no. 9, September 1978, p.33.
57. *Ibid.* p.35.
58. *The Ohio Plain Dealer*, 2—7 August 1981.
59. *CEGB Statistical Yearbook 1982—83*, p.9.
60. The 20 per cent figure is from N. Highton, *op. cit.* (note 33). Other estimates for retro-fitting are slightly higher than this.
61. See page 23.
62. UK Review Group on Acid Rain, *Acid Deposition in the United Kingdom*, Stevenage: Warren Spring Laboratory 1984.
63. Five-year research project, co-sponsored by the CEGB and NCB to the tune of £500,000 each year, run under the auspices

of the Royal Society.

64. *Royal Commission on Environmental Pollution, Tenth Report*, London: HMSO 1984, p.145.

65. Verbal communication from John Kerrie, Lieutenant-Governor of Massachussetts, during a fact-finding tour on acidification in January 1984.

66. See page 96.

67. United Nations Economic Commission for Europe, *Report on Effects of Sulphur Compounds on Soil, Ground Water and Vegetation*, Geneva: 1982, p.11.

68. *Report and Background Papers, 1982 Stockholm Conference on the Acidification of the Environment, Expert Meeting 1*, Stockholm: National Swedish Environment Protection Board 1982, p.15.

69. L. Ginjaar, 'Overall conclusions of the symposium', *Acid Deposition: A Challenge for Europe,* Karlsruhe: EEC 1983, pp.346-7.

70. M.K. Tolba, *The State of the World Environment Report 1983*, Geneva: United Nations Environment Programme 1983, p.6.

71. *CEGB, Annual Report and Accounts 1982—83*, p.26.

72. NRC, *Acid Deposition: Atmospheric Processes in Eastern North America*, Washington: National Academy Press 1983, Introduction.

73. From notes taken at the time. However, the report is confirmed in M.W. Holdgate, 'The ecological effects of deposited sulphur and nitrogen compounds', *Acid Deposition: A Challenge for Europe*, Karlsruhe: EEC 1983, p.339.

74. S. Beilke, 'Report on session: origin, transport, conversion and deposition of air pollutants', *Acid Deposition: A Challenge for Europe*, Karlsruhe: EEC 1983, p.314.

75. From notes taken at the time. Proceedings not published.

76. See page 95.

77. See page 96.

78. See page 97.

79. See page 94.

80. *New Statesman*, 23 September 1983, and *The Times*, 29 October 1983.

81. *Canada News Facts*, Canadian High Commission, 29 October 1982.

82. Speech, 25 February 1983.

83. *Canada News Facts*, Canadian High Commission, 16—23

February 1983.

84. *New York Times*, 8 June 1982.

85. *New Scientist*, 24 June 1982.

86. *Acid News*, no. 4, June 1983, p.9.

87. W. Dickson, 'Water acidification − effects and countermeasures', *Ecological Effects of Acid Deposition*, Stockholm: National Swedish Environment Protection Board 1982, p.273.

88. From Harriman and Morrison, 'Forestry, fisheries, and acid rain in Scotland', quoted in V. Carroll, *Acid Deposition in Scotland and Northern England*, 1983, unpublished research document for Greenpeace, p.19.

89. *Ibid.* p.20.

4. Acid countries

1. W. Dickson, 'Water acidification−effects and countermeasures', *Ecological Effects of Acid Deposition*, Stockholm: National Swedish Environment Protection Board 1982, p.269.

2. Environment '82 Committee, *Acidification Today and Tomorrow*, Stockholm: Swedish Ministry of Agriculture 1982, p.50.

3. Dickson, *op. cit.* p.269.

4. Verbal information from Swedish Embassy, London. Confirmed by *Statens livsmedelsverks kungörelse med allmäna råd för konsumtion av fisk och om saluförbud för fisk m m fråan vissa våattenomraden*, National Swedish Food Administration, 25 May 1983.

5. Verbal communication from Swedish NGO Secretariat on Acid Rain, Stockholm; also see Environment '82 Committee, *op. cit.* p.184.

6. M. Segnestam, *The Problems Caused by Acid Rain in Scandinavia*, European Environmental Bureau Seminar on Acid Rain, February 1982, p.4.

7. H. Hultberg, 'Effects of acid deposition on aquatic ecosystems', *Acid Deposition: A Challenge for Europe*, Karlsruhe: EEC 1983, p.177.

8. G. Lean, 'Awful secret of the lake that died', *The Observer*, 19 September 1982.

9. US Senate Congressional Records, 22 October 1981.

10. Personal communication from Swedish NGO Secretariat on Acid Rain.

11. OECD, *The Costs and Benefits of Sulphur Oxide Control*, Paris: 1981, p.74.
12. B. Assarsson, 'The Swedish sulphur control policy', *Costs of Coal Pollution Abatement*, Paris: OECD 1983, p.103.
13. *Acid News*, no. 4, June 1983, p.9.
14. Dickson, *op. cit.* p.269.
15. I.P. Muniz, 'The effects of acidification on Norwegian freshwater ecosystems', *Ecological Effects of Acid Deposition*, Stockholm: National Swedish Environment Protection Board 1982, p.300.
16. *Ibid.* p.317.
17. *Ibid.* p.304.
18. I.P. Muniz and H. Leivestad, 'Acidification−effects on freshwater fish', *Ecological Impact of Acid Precipitation*, Oslo: SNSF 1980, p.91.
19. *Ibid.* p.92.
20. Erik Lynne, Norwegian Director General, Ministry of Environment, 'Acid rain: international perspectives', speech given August 1982, p.2.
21. J. Karevoll, 'The rain that kills', *The Norseman*, no. 1, 1983, p.13.
22. Lynne, *op. cit.* pp.2−3.
23. M. K. Tolba, *The State of the World Environment Report 1983*, Geneva: UNEP 1983, p.8.
24. *New Scientist*, 27 October 1983.
25. Cf. *Guardian*, 11 November 1983: 'according to the Census of Woodland and Trees ... the total woodland area in England has risen to 948,000 hectares' [3,660 square miles].
26. B. Moynahan, 'The rain that kills', *Sunday Times*, 5 June 1983, p.13.
27. European Communities R & D Information No. 23, *Acid Rain−A Challenge for Europe*, Brussels: 1983, p.5.
28. Herr Muntingh, *Information Note on Acid Rain*, Brussels: European Parliament Committee on the Environment, Public Health and Consumer Protection 1983, pp.4−5.
29. This is a simplification of Ulrich's theories. See Ulrich, 'Effects of accumulation of air pollutants in forest ecosystems', *Acid Deposition: A Challenge for Europe*, Karlsruhe: EEC 1983, pp.127−43.
30. EEC, *Proposal for a Council Directive on Limiting Emissions from Large Combustion Plants*, 15 December 1983.

31. *ENDS Report 104*, September 1983, p.11.
32. *Daily Telegraph*, 3 January 1984. Figures checked with DoE March 1984: the £600,000 refers to financial year 1983−84; the £1 million to 1984−85.
33. *The Scotsman*, 17 August 1983.
34. CEGB Press Release, 28 October 1982.
35. UK Review Group on Acid Rain, *Acid Deposition in the UK*, Stevenage: Warren Spring Laboratory 1984.
36. Friends of the Earth, press release, January 1984.
37. G. Lean, 'British lakes poisoned by acid rain', *The Observer*, 18 September 1983.
38. B. Morrison, 'Acid waters in the UK', lecture given at the BANC/UCL Conference on Acidification and Nature Conservation, University College, 16 December 1983. Information taken from notes.
39. *The Galloway*, 25 November 1982.
40. V. Carroll, *Acid Deposition in Scotland and Northern England*, 1983, unpublished research document for Greenpeace, pp.13 and 20.
41. W.R. Howells, 'The effects of acid precipitation and land use on water quality and ecology in Wales and the implications for the Authority', Welsh Water Authority, Powys, 27 April 1983, p.2.
42. G. Lean, 'Acid rain kills fish in Welsh rivers', *The Observer*, 25 September 1983.
43. W.R. Howells, *op. cit.* p.1.
44. *Evening News*, 2 July 1983.
45. Telephone conversation with Welsh Water Authority. No imputation is intended on the Authority's work, which is among the best in the UK as far as acid precipitation research is concerned.
46. D. W. Sutcliffe and T. R. Carrick, *Acid Rain and the River Duddon in Cumbria*, Ambleside: Freshwater Biological Association 1983, pp.2 and 3.
47. F.T. Last, 'Acid rain−a matter for concern?', *Scottish Wildlife*, vol.19, part 2, May 1983, p.13.
48. Personal communication from Friends of the Earth.
49. *Britain's Energy Resources*, London: HMSO 1982, p.3.
50. I. Tripet and P. Wiederkehr, *Étude du problème des précipitations acides en Suisse*, Lausanne: École Polytechnique Fédérale de Lausanne 1983, p.65.

51. A. Rebsdorf, 'Are Danish lakes threatened by acid rain?' *Ecological Effects of Acid Deposition*, Stockholm: National Swedish Environment Protection Board 1982, p.297.

52. Environmental Resources Ltd, *Acid Rain*, London: Graham and Trotman 1983, p.75.

53. *New Scientist*, 12 January 1984.

54. OECD, *Costs of Coal Pollution Abatement*, Paris: 1983, pp.95 and 97.

55. *New Scientist*, 13 October 1983.

56. Canadian Government Press Release, 22 March 1984.

57. French Ministry of Environment, *Le Reseau de Mesure des pluies acides en France*, Neuilly: 6 February 1984, pp.5−6.

58. French Government Press Release, 22 February 1984.

59. *New Scientist*, 1 March 1984.

60. S. Postel, *Forests in a Fossil-Fuel World*, Worldwatch March 1984.

61. T. Heyse, *Acid Precipitation: The Situation in Belgium*, 1983, unpublished Greenpeace research document.

62. Muntingh, *op. cit.* p.5.

63. *International Herald Tribune*, 4 January 1984.

64. Muntingh, *op. cit.* p.5.

65. S. Postel, *Forests in a Fossil-Fuel World*, Worldwatch March 1984.

66. J. Cooke, 'Acid rain turnabout', *Earthscan*, 1983, p.1.

67. Muntingh, *op. cit.* p.5.

68. Environment '82 Committee, *Acidification Today and Tomorrow*, Stockholm: Swedish Ministry of Agriculture 1982, p.69.

69. Verbal communication from Bo Landin.

70. *Guardian*, 9 February 1984.

71. H.C. Martin, 'Acidification of the environment: a Canadian perspective', *Acid Rain: A Challenge for Europe*, Karlsruhe: EEC 1983, p.186.

72. *Ibid.* pp.186−9.

73. US Senate Congressional Record, 22 October 1981.

74. *Acid Rain*, Ottawa: Department of Fisheries and Oceans, 1982, p.8.

75. Martin, *op. cit.* pp.187−9.

76. Martin, *op. cit.* p.188.

77. M. Munro, 'The silent peril', *Ottawa Citizen*, reprinted by Environment Canada, n.d., pp.1−2.

78. P. Weller, 'Industry and acid rain: the Canadian corporate response', *Alternatives*, Winter 1983, pp.21—2.
79. Martin, *op. cit.* p.189.
80. It is generally accepted that half the sulphur dioxide in Canadian airspace comes from the USA. See Martin, *op. cit.*, p.189.
81. The Canadians offered to spend $10,000 million if the Americans spent a similar amount. *New Scientist*, 3 February 1983.
82. *New Scientist*, 9 February 1984.
83. R. Hoyle, 'The silent scourge', *Time*, 8 November 1982.
84. H.W. Vogelman, 'Catastrophe on Camel's Hump', *Natural History*, November 1982.
85. 'Catastrophe on Camel's Hump, continued', *Natural History*, August 1983.
86. E. Cowling, 'A historical resumé of progress in scientific and public understanding of acid precipitation and its biological consequences', 1980, quoted in *Acidification Today and Tomorrow*, p.68 (note 2).
87. National Wildlife Federation, *Acid Rain*, 1982, pp.5—6.
88. Environment Canada, *Downwind: the Acid Rain Story*, Ottawa: 1982, p.15.
89. Speech, 11 April 1983.
90. J. Kraus, 'Legal approaches to the control of acid rain', *Alternatives*, vol.11, no. 2, Winter 1983, p.19.
91. Telephone conversation with the Campaign for Lead-free Air. The catalysts are discussed in *Royal Commission on Environmental Pollution, Tenth Report*, London: HMSO 1984, p.134.
92. Kraus, *op. cit.* p.30.
93. *Ibid.* p.33.
94. A.J. Apling, 'Current understanding of the technical possibilities for reducing emissions from motor cars', *Acid Deposition: A Challenge for Europe*, Karlsruhe: EEC 1983, pp.260 and 262.
95. Hoyle, *op. cit.* p.43.
96. Personal communication at the Karlsruhe symposium. William Ruckelshaus, director of the Environmental Protection Agency, was later quoted as saying: 'Members of the US cabinet have brought different perspectives to the issue' (*Evening Argus*, 24 October 1983).
97. E. Kemf, 'Acid rain in the Arctic', *World Wildlife News*, no. 27,

January−February 1984, p1.

98. IUCN Features Service, 'Acidification threatens southern hemisphere', 15 December 1983, p.2.

99. M. Margolis, 'Life under a cloud−looking for the silver in a sulfur lining', *Christian Science Monitor*, 30 May 1983.

100. *The Times*, 14 May 1983.

101. IUCN Features Service, *op. cit.* pp.1−2.

102. *Ibid. p.2.*

Appendix A: Research into acid rain

1. A. Littler, 'Flue-gas washing at power stations in the UK 1933−1976', London: Central Electricity Generating Board, July 1976.

2. Much of the information in this appendix is from E.B. Cowling, 'Acid precipitation in historical perspective', *Environmental Science and Technology*, vol. 16, no. 2, 1982.

Index